Metagenomics for Microbiology

Metagenomics for Microbiology

Edited by

Jacques Izard

The Forsyth Institute, Cambridge,
Massachusetts, USA

Maria C. Rivera

Virginia Commonwealth University,
Center for the Study of Biological Complexity,
Richmond, Virginia, USA

AMSTERDAM • BOSTON • HEIDELBERG • LONDON
NEW YORK • OXFORD • PARIS • SAN DIEGO
SAN FRANCISCO • SINGAPORE • SYDNEY • TOKYO

Academic Press is an imprint of Elsevier

Academic Press is an imprint of Elsevier
32 Jamestown Road, London NW1 7BY, UK
525 B Street, Suite 1800, San Diego, CA 92101-4495, USA
225 Wyman Street, Waltham, MA 02451, USA
The Boulevard, Langford Lane, Kidlington, Oxford OX5 1GB, UK

British Library Cataloguing-in-Publication Data
A catalogue record for this book is available from the British Library.

Library of Congress Cataloging-in-Publication Data
A catalog record for this book is available from the Library of Congress.

ISBN: 978-0-12-410472-3

For Information on all Academic Press publications
visit our website at http://store.elsevier.com/

CONTENTS

LIST OF CONTRIBUTORS

Nadim J. Ajami
Department of Molecular Virology and Microbiology, The Alkek Center for Metagenomics and Microbiome Research, Baylor College of Medicine, Houston, TX, USA; Metanome Inc., Houston, TX, USA

Mathieu Almeida
Center for Bioinformatics and Computational Biology, University of Maryland, College Park, MD, USA

Brett Bowman
Pacific Biosciences, Menlo Park, CA, USA

Erika del Castillo
Department of Microbiology, The Forsyth Institute, Cambridge, MA; Harvard School of Dental Medicine, Boston, MA, USA

Yong-Joon Cho
ChunLab, Inc. Seoul National University, Seoul, Korea

Georg K. Gerber
Department of Pathology, Brigham and Women's Hospital, Harvard Medical School, Boston, MA, USA

Shaomei He
Department of Bacteriology and Department of Geoscience, University of Wisconsin-Madison, Madison, WI, USA

Jacques Izard
Department of Microbiology, The Forsyth Institute, Cambridge, MA; Harvard School of Dental Medicine, Boston, MA, USA

Mincheol Kim
School of Biological Sciences, Seoul National University, Seoul, Korea

Jonas Korlach
Pacific Biosciences, Menlo Park, CA, USA

Patricio S. La Rosa
Predictive Analytics, Monsanto, St. Louis, MI, USA

Joseph F. Petrosino
Department of Molecular Virology and Microbiology, The Alkek Center for Metagenomics and Microbiome Research, Baylor College of Medicine, Houston, TX, USA; Metanome Inc., Houston, TX, USA

Mihai Pop
Department of Computer Science; Center for Bioinformatics and Computational Biology, University of Maryland, College Park, MD, USA

Maria C. Rivera
Department of Biology, Center for the Study of Biological Complexity, Virginia Commonwealth University, Richmond, VA, USA

Matthias Scholz
Centre for Integrative Biology, University of Trento, Trento, Italy

Nicola Segata
Centre for Integrative Biology, University of Trento, Trento, Italy

William D. Shannon
Department of Medicine, Washington University School of Medicine, St. Louis, MI, USA; BioRankings, LLC, St. Louis, MI, USA

Erica Sodergren
The Jackson Laboratory for Genomic Medicine, Farmington, CT, USA

Adrian Tett
Centre for Integrative Biology, University of Trento, Trento, Italy

George Weinstock
The Jackson Laboratory for Genomic Medicine, Farmington, CT, USA

Yanjiao Zhou
The Genome Institute, Washington University School of Medicine, St. Louis, MI, USA; Department of Pediatrics, Washington University School of Medicine, St. Louis, MI, USA

PREFACE

It is well known that only a small fraction of extant microbial life has been identified. Metagenomics, the direct sequencing and characterization of genes and genomes present in complex microbial ecosystems (e.g., metagenomes), has revolutionized the practice of microbiology by bypassing the hurdle of pure culture isolation. Metagenomics shows promise of advancing our understanding of the diversity, function, and evolution of the uncultivated majority.

Metagenomics as a field arose in the 1990s after the application of molecular biology techniques to genomic material directly extracted from microbial assemblages present in diverse habitats, including the human body. The application of metagenomic approaches allows for the acquisition of genetic/genomic information from the viruses, bacteria, archaea, fungi, and protists forming complex assemblages. The field of metagenomics addresses the fundamental questions of which microbes are present and what their genes are potentially doing.

In the mid-2000s, the availability of high-throughput or next-generation sequencing technologies propelled the field by lowering the monetary and time constraints imposed by traditional DNA sequencing technologies. These advances have allowed the scientific community to examine the microbiome of diverse environments/habitats, follow spatial and temporal changes in community structure, and study the response of the communities to treatment or environmental modifications.

In 2012, the publication of the large-scale characterization of the microbiome of healthy adults created high expectations about the influence of the microbiota in human health and disease. With the publication of the results of the Human Microbiome Project, metagenomics has emerged as a major research area in microbiology, particularly, when it comes to the characterization of the role of microbiota in complex disorders, such as obesity.

With contributions by leading researchers in the field, we provide a series of chapters describing best practices for the collection and analysis of metagenomic data, as well as the promises and challenges of the field. The chapters have been dedicated to different aspects of metagenomics. Chapter 1 provides an end-to-end overview of the metagenomic pipeline and its challenges. Chapter 2 showcases SMRT, one of the third-generation sequencing platforms, and its use in metagenomics. As high abundance of ribosomal RNA (rRNA) transcripts is a major hurtle for the application of transcriptomics to microbial communities, Chapter 3 describes methodology that can reduce the "noise" rRNA imposes on this type of studies. Chapters 4 and 5 showcase some of the computational approaches that are used to analyze the whole-community metagenome sequence data and available software, and highlight future research directions. The statistical challenges and solutions for cross-sectional and longitudinal data sets are explored in Chapters 6 and 7, respectively. Chapter 8 presents a historical perspective of the microbiome studies, the societal impact of microbial communities, and the challenges ahead for metagenomics, while advances in virome studies are explored in Chapter 9. A perspective on the current efforts, challenges, and the future of metagenomic is presented in Chapter 10.

This book is intended for researchers, teachers, students, and the citizen scientists contemplating performing microbial metagenomics studies. For microbiologists generating metagenomic next-generation sequencing data, the book will provide an introduction and support to the computational and statistical specifics of the data. For the statisticians and computational scientist contemplating working with metagenomic data, it will provide some of the initial background needed. For the community, in general, it will provide the basis for further investigation of this transformative and fascinating field.

We would like to thank all authors for their contributions. We need to acknowledge the public and private funding entities that made this technological and conceptual advance a possibility, as well as the researchers and consortia that broke the grounds for those innovations to flourish. Last, we would like to thank Elsevier for the short book format and allowing a more focused and didactic approach.

Jacques Izard
Maria C. Rivera

CHAPTER 1

Steps in Metagenomics: Let's Avoid Garbage in and Garbage Out

Jacques Izard

WHY METAGENOMICS?

Is metagenomics a revolution or a new fad? Metagenomics is tightly associated with the availability of next-generation sequencing in all its implementations. The key feature of these new technologies, moving beyond the Sanger-based DNA sequencing approach, is the depth of nucleotide sequencing per sample.[1] Knowing much more about a sample changes the traditional paradigms of "What is the most abundant?" or "What is the most significant?" to "What is present and potentially significant that might influence the situation and outcome?"

Let's take the case of identifying proper biomarkers of disease state in the context of chronic disease prevention. Prevention has been deemed as a viable option to avert human chronic diseases and to curb health-care management costs.[2] The actual implementation of any effective preventive measures has proven to be rather difficult. In addition to the typically poor compliance of the general public, the vagueness of the successful validation of habit modification on the long-term risk, points to the need of defining new biomarkers of disease state.

Scientists and the public are accepting the fact that humans are super-organisms, harboring both a human genome and a microbial genome, the latter being much bigger in size and diversity, and key for the health of individuals.[3,4] It is time to investigate the intricate relationship between humans and their associated microbiota and how this relationship modulates or affects both partners.[5] These remarks can be expanded to the animal and plant kingdoms, and holistically to the Earth's biome. By its nature, the evolution and function of all the Earth's biomes are influenced by a myriad of interactions between and among microbes (planktonic, in biofilms or host associated) and the surrounding physical environment.

Metagenomics for Microbiology. http://dx.doi.org/10.1016/B978-0-12-410472-3.00001-4

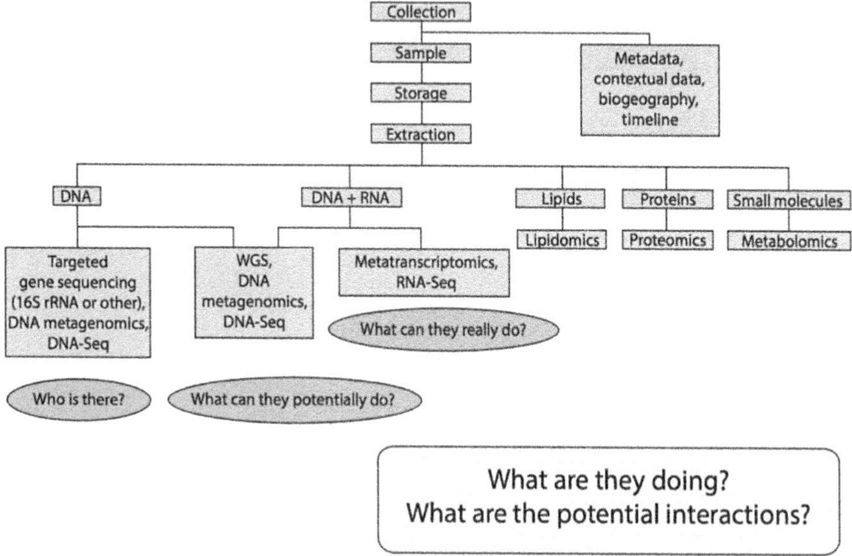

Fig. 1.1. Metagenomic analysis process and some of the overarching questions that can be answered by the different methodologies.

The general definition of metagenomics is the cultivation-independent analysis of the genetic information of the collective genomes of the microbes within a given environment based on its sampling. It focuses on the collection of genetic information through sequencing that can target DNA, RNA, or both. The subsequent analyses can be solely focused on sequence conservation, phylogenetic, phylogenomic, function, or genetic diversity representation including yet-to-be annotated genes. The diversity of hypotheses, questions, and goals to be accomplished is endless. The primary design is based on the nature of the material to be analyzed and its primary function (Figure 1.1).

IT ALL STARTS WITH THE STUDY DESIGN

The goal is not to tell you how to do your science but to emphasize some aspects of study design that need careful attention because of the characteristics of the methodologies used in metagenomic studies. It begins by identifying the primary objective of the metagenomics project. What is the main scientific question you are trying to answer? More than one hypothesis can be tested depending on the scope of the experiment and the amount of associated data, or metadata, that you collect and use for your subsequent analyses.

The high-dimensionality characteristic of the metagenomics datasets is challenging and is revolutionizing microbiology analytical methodology. What is meant by high-dimensional dataset? Let's take as an example the Human Microbiome Project (HMP) 16S ribosomal RNA (rRNA)-based characterization of 10 sites from the digestive tract of 200 individuals. Such analysis required the collection of over 2000 samples, generating approximately 23 million high-quality sequence reads that were assigned to 674 taxonomic clades with their respective relative abundance per taxonomic level (e.g., from phylum to genus). For example, for the genus *Pyramidobacter*, the database stores the relative abundance at each taxonomic level, from the phylum (e.g., "Bacteria|Synergistetes"), the most inclusive taxonomic level, to the genus (e.g., "Bacteria|Synergistetes|Synergistia|Synergistales|Synergistaceae|*Pyramidobacter*"), the least inclusive taxonomic level, and all the taxonomic levels between the two.[6] From the same study, four body sites were further analyzed using whole metagenome shotgun (WMS) sequencing from approximately 100 individuals, generating a trillion nucleotides.[6] Another example can be extracted from the work of Giannoukos et al.[7] while developing rRNA depletion methodology for fecal samples. They obtained over 100,000 reads per sample.[7] In each example, each sample has a tremendous amount of genotypic and phenotypic information in addition to the metadata (e.g., age, sex, race, and others). In addition to the nucleotide data, information about other molecules (e.g., lipids, proteins, and metabolites) can be collected; increasing the complexity and multidimensionality of the dataset. The type of data collected will determine the type of analyses performed. These analyses can help answer questions such as: "What are the organisms present?", "What can these organisms potentially do?",' "What is their metabolic capability?", and "How do they influence the host?" (Figure 1.1). Planning the structure of samples and metadata acquisition as well as the analysis pipeline to be used, prior to the start of the experiment, will avoid bottlenecks and optimize utilization of funds.

During the study design phase, investigators need to take into consideration the ethical and legal issues related to metagenomics data collection and analysis. Some of the constrains of metagenomics studies utilizing human subjects include Institutional Review Boards, informed consent, and other issues related to the protection of the identifiable health information of the human subjects (e.g., HIPAA Privacy Rule in the United States). For examples of consent documentation and standard operating procedures, the National Institutes of Health HMP has made those document public and available online (http://www.hmpdacc.org).[8] It is essential for the consent procedures to accurately state what data will be gathered, how it will be used, and how it will be stored. All efforts should be made to secure information and confidentiality of the genetic material and associated data over time. This includes both the physical storage of the information, data deposition and data sharing, even when the samples are de-identified. For environmental samples, having the right of access and sampling permits is critical as geolocation is now required with the sample data submission to repository.

It is important to point out that any samples collected from a host will contain a significant amount of the host genetic material. The potential contamination of samples with the host genetic material adds to the complexity of the metagenomics studies, and sophisticated computational pipelines for the removal of the contaminating reads are essential to generate meaningful conclusions and, in the case of human subjects, to protect the privacy and confidentiality of the sample donor. Figure 1.2 shows the impact of human "contamination" on the amount and quality of the data collected using shotgun sequencing of human samples from 16 different body sites.[8] When working with different models, it should be noted that the genome of a brown rat is not that much smaller than that of a human (over 3 billion base pairs), and that the corn genome is over 2 billion base pairs. Although protists and fungi are much smaller, their genomes are still composed of few million base pairs. The knowledge of your biological system of interest will be critical to optimize the study design.

HAVE A STATISTICAL ANALYSIS PLAN IN PLACE BEFORE STARTING

Planning for statistical analysis should be an integral part of the study design. Although many experimental designs can be performed in metagenomics project, there is no single path to a successful strategy.

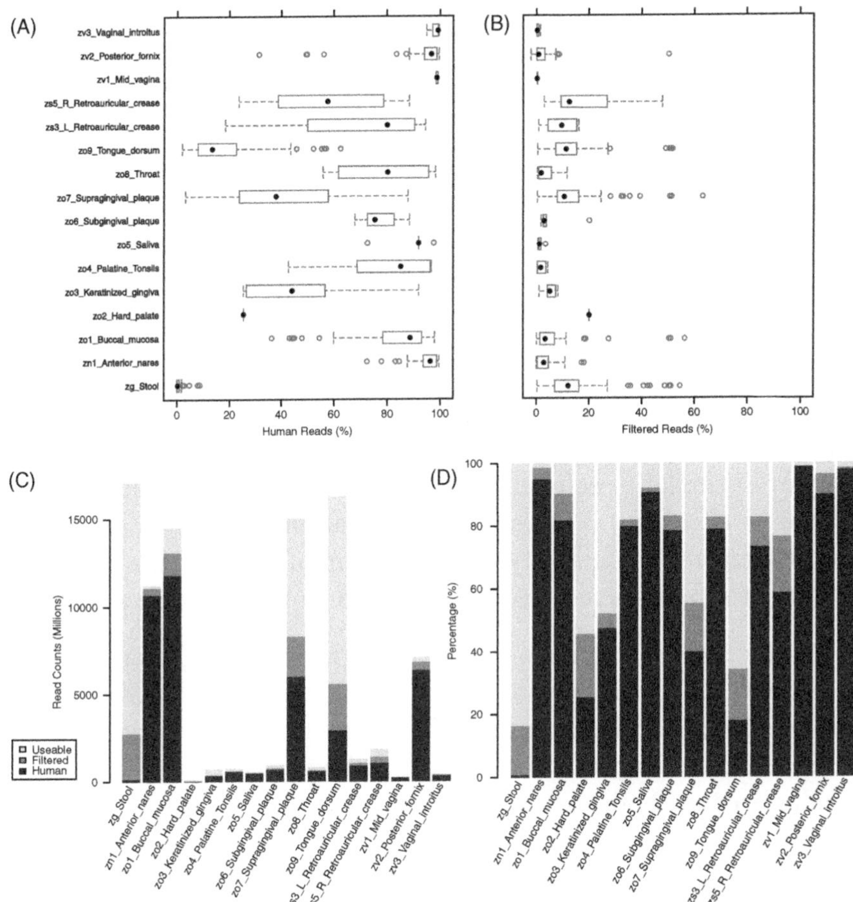

Fig. 1.2. Impact of quality and human filtering on shotgun metagenomic dataset. Thorough quality filtering and removal of reads resulting from human DNA contamination was performed on all shotgun metagenomic data of the Human Microbiome Project (average of 13 Gb/sample). The variation in fraction of reads per sample removed across the 18 body sites is shown by (A) boxplots for % of human and of (B) quality filtered reads. (C) Total amount of usable data (white) per site significantly varied because of (i) the different number of samples per site, (ii) the differential impact of human contamination (dark gray), and (iii) the differential impact of quality filtering (light gray). (D) Summary view of the usable fractions versus human and quality filtered data, per body site. (Reprinted by permission from Macmillan Publishers Ltd.[8])

While using metagenomic or metatranscriptomic approaches, it is essential to refer to the specific needs of each experiment.

The statistical analysis plan should take into account the characteristics of the experiment (in human studies, this would be the inclusion and exclusions criteria), the rate of sample acquisition (this would include the rate of human subject recruitment that will determine if you are working with one or more batch of datasets), the descriptive objectives, testable hypotheses, the statistical methods that might be stand alone or imbedded in bioinformatics tools or pipelines, etc. One of the direct advantages of planning ahead is that when you have the data in hand, you'll have a strategy in place to start the analysis. This is critical as next-generation sequencing provides a tremendous amount of data and you want to remain focused on your primary objective(s). After the accomplishment of your primary objective(s), exploratory analyses and additional hypotheses investigation or formulation is always a possibility.

The most basic question about the research plan should be "Are enough samples being collected from each site or from enough subjects to make meaningful conclusions?" To properly assess the degree of similarity or dissimilarity between bacterial communities, a measurable difference, or effect size, is necessary. In general, the smaller the effect size and the greater the variability within a group of samples, the larger the number of samples is required to achieve adequate statistical power.

For determining sample size for experiments using metagenomic taxonomic data, the work derived from the HMP provided the first available calculation and software package[9] (see chapter 6 by La Rosa and colleagues). For metagenomics and metatranscriptomics, standardized methods to assess the number of subjects (or independent samples) and reads are yet to be developed. If you are planning to use both a 16S rRNA gene-targeted approach and whole-metagenome shotgun sequencing, a two-stage experimental design is an option to focus on a subset of samples.[10]

The complexity of your sample will greatly influence the depth of sequence coverage in WMS and metatranscriptomics sequencing projects. As mentioned above, host genomic information can represent a significant amount of genetic data obtained through next-generation sequencing approaches, and this information should be part of an optimized study design.

If the complexity of the sample is low (as determined by more traditional methods), you may be able to estimate the depth of sequencing coverage needed, in order to sample the whole metagenome. Although each next-generation sequencing platform has its unique biases and associated errors (an issue not restricted to next-generation sequencing),

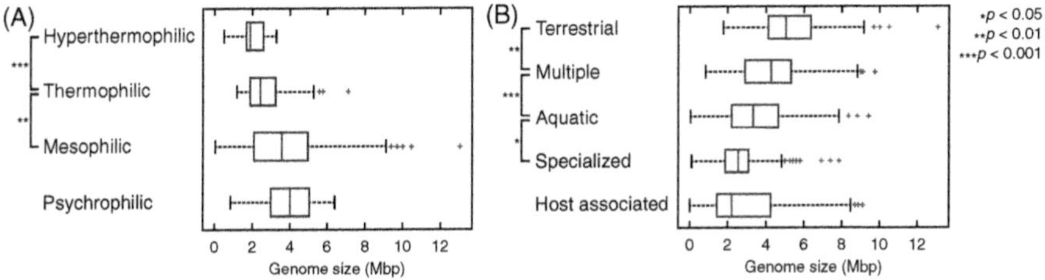

Fig. 1.3. Distribution of genome size based on temperature and habitat. (A) Distribution of genome sizes among prokaryotes with different growth temperature ranges. The differences in genome size between mesophiles, thermophiles, and hyperthermophiles are significant (Wilcoxon rank-sum test, P < *1.9* × *10*$^{-5}$ *and* P < *7.9* × *10*$^{-3}$ *for mesophiles–thermophiles and thermophiles–hyperthermophiles, respectively), but not between psychrophiles and mesophiles (Wilcoxon rank-sum test,* P = *0.082). (B) Distribution of genome sizes among different habitats. Habitats are ordered according to environmental variability from unvarying (host associated) to the most variable environment (terrestrial). The distributions of genome sizes differ between habitats (Wilcoxon rank-sum test,* P < *0.018,* P < *0.0005,* P < *0.0028, for specialized-aquatic, aquatic-multiple, and multiple-terrestrial, respectively), with the exception of host-associated habitats (Wilcoxon rank-sum test,* P = *0.67, for comparison between host-associated and specialized). The red vertical marks are the medians, the edges of the box are the 25th and 75th percentiles, the whiskers extend to the most extreme data points not considered outliers (99% of all data if the data are normally distributed), and outliers are individually plotted as red crosses.* Reprinted by permission from Oxford University Press.[16]

metagenomic analyses assume that the reads are sampled randomly, independently, and evenly distributed across all the genomes in the metagenome.[11,12] To calculate the coverage, you need to know the amount of material (nucleotide amount) you are using and the size of the genomes or an average size for that environment. Figure 1.3 provides an overview of expected genome size in prokaryotes that can be complemented by other resources providing the exact information on specific genomes.[13–16] The correlation between G+C content and chromosome size can be positive, negative, or not significant depending on the clade from kingdom to species.[15] To our advantage, most chromosomes within a species have a similar pattern of correlation between G+C content and chromosome size; however, outliers are common.[15]

Longitudinal studies present their own challenges and can be independently analyzed at each time point, along the timeline as well as across body sites[17,18] (see chapter 7 in this book). When feasible, the collection of the metadata in between the time points is also critical in understanding the dynamic signatures of microbial population modification.

Pooling the samples might seem to be a good strategy to reduce cost and reduce sample variation. However, this approach loses all of the low genetic representation and the ability to make inferences about the microbial population.

You might not find a metagenomic dataset to help or guide you in the experimental design phase. Instead, previous results using other molecular techniques or culture-based methods might be an alternative source of help in the design. If you were looking at the same question with a more traditional method, you should have enough samples to detect differences if they are present.

METADATA IS NEEDED TO PROVIDE CONTEXT TO THE ANALYSIS

Critical to any metagenomic study is the quality and extent of the contextual metadata. Metadata is what will enhance your analysis beyond the most obvious evidence. It provides context to the experiments and allows for meaningful comparisons between studies, while deepening our understanding of the dataset. With a greater depth of information, a broader knowledge of the "environmental factors" is needed. Although not the focus of an experiment, seemingly extraneous data may become important. For example, information on the source of carbon for microbial metabolism might be later identified as a confounding variable in an experiment. It can be as simple as the source of sugar intake for a subject or the nature of the pollutant for a soil sample.

The information about the sample location or its relative position to other samples can be included in the analyses. The concept of biogeography goes beyond the description of environmental features that influence the spatial distribution of the microorganisms. It aims to understand the metabolic processes within the microbes' own niche and their relationships with other biological niches. The niche might be the different sites in the oral cavity, along the digestive tube, or in the skin.[19–21] Large-scale data visualization and analysis tools have been created to help us better understand these positional aspects[22].

As we are discovering the microbiome as an interdependent organ of any biological system,[5] we may need to redefine what are the best associated data to collect along with the genomic sample. Although blood analyses might reflect the systemic inflammation of a human subject, the levels of air particles less than 2.5 μm in diameter ($PM_{2.5}$) that the subject is exposed to might contribute to the severity of their asthma, modifying the microbiome, which, in turn, can modify the responsiveness to medication.[23,24] In longitudinal datasets, seasons and length of the day have been shown to influence the ocean microbiome.[25]

Defining or re-defining the phenotype of interest might have a crucial importance. Because the phenotype is the results of the interaction between the genotype and the organism's environment in all its complexity, including the microbiome, we are required to renew our attention to the granularity of the defined phenotype. From the macro to the molecular scale, new considerations that were previously neglected because of the lack of significance might be at play when scrutinized with a different sliver or window of observation. Guidelines for data organization and naming standardization are already in place and are being improved upon, as described below.

SAMPLING: THE BASIS OF GOOD RESULTS

Although the technology of the sequencing platforms has evolved, they all focus on sequencing the nucleic acids, either DNA or RNA. The source of the microbiome sample greatly varies, from the environment, plants, insects, and animals to humans. The published data on environmental samples have been as diverse as soil, hot springs, seawater, air, as well as home and hospital surfaces. For plants, the associated microbiome above and below the ground has been studied. In insects, animals, and humans, multiple body sites have been investigated. In many of the subsequent steps, the hypothesis involved, the goals of the project, the available facilities and personnel, and the available funds play a role in the decision matrix.

Contamination will be detected as an integral component of the sample because of the depth of the data being acquired. Only a few years ago, understanding microbial diversity often led the investigator to do a series of cloning experiments resulting in the identification of approximately 100 randomly selected organisms per sample. Later, the availability of microarrays allowed the identification of few hundreds of organisms per sample. More recently, by using targeted 16S rRNA gene next-generation sequencing, tens of thousands of organisms can be identified per sample.[1,26] It is recommended to examine each step in the context of potential inadvertent contamination by nucleic material or potential inhibitor for downstream applications. This is particularly applicable to tools that are reused, where proper cleaning and sterilization procedures are essential. The following guidelines are simple ways to increase the quality of sample preparation. Not talking over a biological sample or wearing a facemask would eliminate contamination by the breath. While protecting the sample using gloves, we should not forget that a simple touch of the skin or a surface would contaminate the glove that, in turn, might contaminate the sample itself. Natural DNAses and RNAses may potentially damage the sample. It is most often about applying common sense in the context of the depth of the data to be gathered. In other words, if you want to know the microbiome of the banana peel on the plant but you drop the banana in the field, you are going to also learn about the microbiome of that square of earth as well as that of the fruit.

The proper sampling protocol is essential to a successful metagenomics study, since the accurate identification of many organisms depends on the collection and handling of the sample. Defining the geographical location or the specific body site, surface, depth volume, or quantity to be collected are necessary for sampling standardization. When possible, keep the samples concentrated and process them for immediate storage. Consistency in all aspects will both preserve the quality of the sample and limit the batch effect during the analysis, enhancing the signal of interest. Protecting the samples against the element (wind, sun, etc.) sounds to be a good advice, but keep in mind that sample desiccation is a common problem when working with small samples.

Analyzing true and technical replicates of a sample and assessing whether observed differences are statistically significant are a good practice. True replicates, when the same site is sampled more than once, are rarely done in metagenomics study as the sensitivity of the technique may easily show differences when sampling a site multiple times because of the biological organization of the site.[27] Technical replicates, when the sample is split for processing, are easy to perform for reassurance.[28,29]

SAMPLE STORAGE

Storage and sampling are tightly linked issues. It is not always possible to have a freezer or an expert on location when the sample is collected. Solutions for these problems affecting the downstream steps need to be identified before starting the study. The nature of the type of sample is too diverse to enter in all the details, but one key question will drive the process: "How much sample do I really need?" The associated questions would be: "Do I need DNA, RNA, proteins, lipids, small molecules, etc., from the same sample?", "Will the sample be used for more than one application, preparation, or extraction?", as well as any other questions related to the present or the future study applications that might be of interest later on.

Many options are available, from immediate extraction to long-term storage in liquid nitrogen. The nature of the sample often dictates what is the best protocol to avoid sample desiccation, denaturation, lysis, degradation, etc. As immediate extraction on site or access to an −80°C freezer is not always an option, alternatives must be developed to preserve the sample, its integrity, and its value for the question(s) at hand. Similarly, for a vaccine, the quality of storage and its consistency might influence the sample quality. Multiple companies are offering sampling kits with fixative but those are rarely validated by comparative analysis. A metagenomic and metatranscriptomic comparison of human stools flash frozen, preserved in ethanol, or in RNA later show that those fixatives are compatible with large-scale self-collection by human subjects in a geographically disseminated cohort.[30]

Sampling cost is often neglected. You might have multiple steps in your process to reach the final storage space, and there is no issue with that. Optimize your process to be the most consistent for each sample or per batch of samples. It should be stressed that whether you work on a large human subject cohort or a large field collection, the cost of personnel, sampling equipment (single use when possible), and transient as well as permanent storage adds up quickly. With the sample collected and in storage, nucleotide extraction will be the next step.

SAMPLE EXTRACTION

The sample input into a metagenomics pipeline can be extremely diverse. The DNA and/or the RNA need to be extracted from the sample prior to any analysis. The type and source of the sample determines the most appropriate extraction protocol. This step, simplified by the availability of nucleic acid extraction kits, is crucial to the success of the analysis, as the quality of extracted DNA and/or RNA influences all subsequent steps. Before selecting the most appropriate extraction protocol, a careful review of the literature and validation of the protocol for your specific sample is recommended. The choice of protocol depends on the DNA or the RNA yield, shearing, removal of contaminants (which could be inhibitory to subsequent steps), and representation of diversity. A compiled list of extraction protocols for different sample sources has been recently published.[31] Some other criteria have to be taken in consideration as described below.

As mentioned above, the source of the sample is very important in the selection of the extraction protocol. A classic example of this is demonstrated by the inhibitory effect of humic acids in enzymatic reactions, such as polymerase chain reaction (PCR), performed using nucleic acids extracted from manure or soil.[32,33] Thus, elimination of humic acids needs to be part of the process, which might be already optimized by a compatible specific kit.

How the sample was preserved also matters. An example is the DNA recovery from formalin-fixed paraffin-embedded tissue, as the tissue is not readily available to traditional protocols.[34,35]

Differences in the structures of bacterial cell walls cause bacterial cell lysis to be more or less efficient.[36,37] The differential efficiency of the lysis can distort the apparent composition of the microbial communities and introduce bias in estimates of relative abundance.[36–39]

Consistency in sample handling and processing is key to avoid batch effect. Training, standard operating procedure, and good quality controls greatly help in minimizing the possibility of batch effect. Nucleic acids extraction automation is a good alternative when sufficient samples are available and the method of extraction has been validated.[40]

Extracting more than one macromolecule at the time is an option. Kits and protocols allow the purification of both DNA and RNA from the same sample, while others go further by recovering proteins as well.[31] An ongoing challenge is to purify other macromolecules from the same sample, which might require a different set of strategies.

Removing the host DNA might improve the quality of your analysis and decrease the cost of the sequencing by requiring magnitude(s) less of reads for the same amount of information. Differential lysis of eukaryotic cells (personal communication, Dr Eva Haenssler) and separation of methylated DNA based on CpG site methylation density between the host and the microbes[41] are the two strategies used by commercial kits. The attempt to decrease host DNA is not only limited to vertebrate hosts but successful contaminant DNA removal have also been performed in plants.[42,43]

CHOOSING THE RIGHT PLATFORM

The cost of sequencing has drastically decreased (Figure 1.4), opening the door to many new investigations that were previously too costly. Although the cost per base of sequencing has decreased, the total cost of a run is still significant because the number of megabases sequenced per run is steadily increasing (Table 1.1). The initial entry cost might be still too high for some pilot projects. Based on those same parameters, traditional

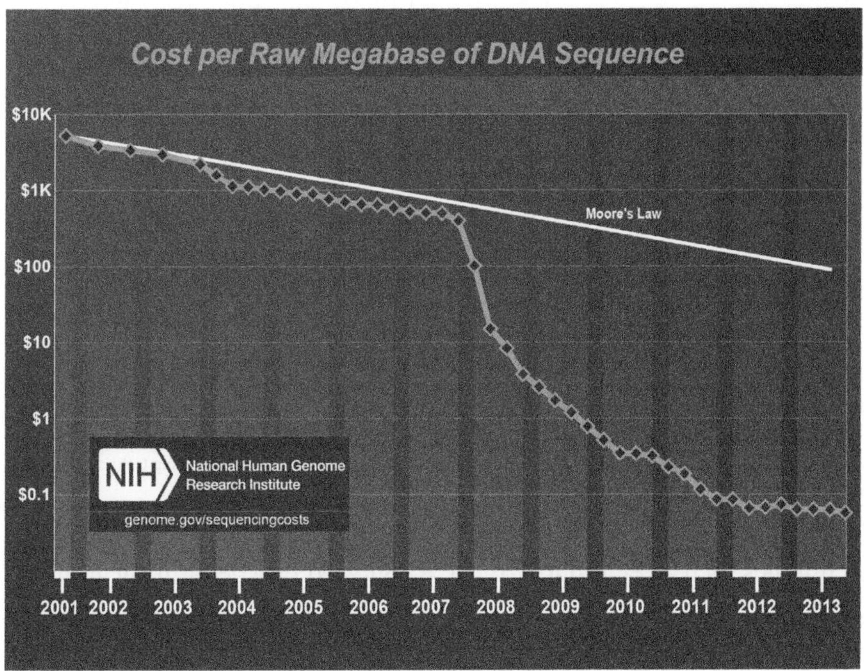

Fig. 1.4. Reduction of the cost of DNA sequencing over time. The white line reflects the Moore's Law pace. The Y axis shows, in logarithmic scale, the cost of sequencing per raw megabase of raw unassembled DNA sequence. The out-pacing of Moore's Law pace matches the availability of the first next-generation sequencing platforms, in 2008, competing with Sanger-based DNA sequencing technology. (Courtesy of the National Human Genome Research Institute.[45])

Table 1.1 Sequencing Platforms and Characteristics Based on Online Manufacturer Technical Specifications.

Sequencer	Read Length (b)[a]	Run Time (h) (d)[b]	Reads Per Run	Yield (b)[a]	Mate Pair Information	Use in Metagenomics
ABI 3730xl[c]	500–900	6–10 h	–	0.05–0.08 Mb	Yes	Not anymore
Roche 454 GS Junior[d]	~400	10 h	~100,000	35 Mb	No	Yes
Roche 454 GS FLX+[d]	~700	23 h	1 million	700 Mb	No	Yes
Illumina MiSeq[d]	~300	5–65 h	25 million	0.3–15 Gb	Yes	Yes
Illumina NextSeq 500	~300	12-30 h	130–400 million	20–120 Gb	Yes	Yes
Illumina HiSeq 2500[d]	~125–150	7 h to 6 d	300 million to 2 billions	10–180 Gb	Yes	Yes
Illumina HiSeq X[d]	~150	<3 d	3 billions	1.6–1.8 Tb	Yes	Not yet
5500 SOLiD[d]	~60	7 d	–	90 Gb	Yes	Yes
5500xl SOLiD[d]	~60	7 d	–	300 Gb	Yes	Yes
Ion PGM system[e]	~200 or ~400	2–4 h or 4–7 h	0.4–0.5 million on 314 chip 2–3 million on 316 chip 4–5.5 million on 318 chip	30–100 Mb on 314 chip 300 Mb to 1 Gb on 316 chip 600 Mb to 2 Gb on 318 chip	No	Yes
PacBio RS II SMRT[e]	4200–8500	0.5–3 h	50,000 per cell	275–375 Mb per cell	No	Yes

[a]b stands for base and its multiple
[b]h: hours; d: days
[c]First-generation DNA sequencing or Sanger-based DNA sequencing technology. ABI 3730xl: Applied Biosystems 3730xl DNA Analyzer (Life Technologies Corporation, Carlsbad, CA).
[d]Second-generation DNA sequencing. Roche 454 GS Junior and Roche 454 GS FLX+ systems from Roche Diagnostics Corporation (454 Life Sciences, Branford, CT). Illumina MiSeq, HiSeq 2500 and HiSeq X from Illumina, Inc. (San Diego, VA). 5500 and 5500xl SOLiD sequencer from Life Technologies Corporation (Carlsbad, CA).
[e]Third-generation DNA sequencing. Ion PGM system from Life Technologies Corporation (Carlsbad, CA). PacBio RS II SMRT system is based on single-molecule, real-time (SMRT) DNA sequencing technology from Pacific Biosciences (Menlo Park, CA).

techniques such as PCR-DGGE (PCR followed by a denaturing gradient gel electrophoresis), cloning experiments followed by Sanger-based DNA sequencing, and microarrays approaches are here to stay.[44,45]

Which sequencing platform to use? Because of the varied nature of scientific studies, there is no single approach that is recommended. Detailed review of the literature, discussions with colleagues and sequencing facilities, cost, availability, turnaround time, and scope of the project will be part of the decision-making process. Let's not forget that the hypothesis and the goal should be the true drivers. Table 1.1 shows the characteristics of the different high-throughput sequencing technologies. Each sequencing platform is characterized by their strength and weaknesses regarding read length, bias in AT- or GC-rich regions and their ability to sequence homopolymers.[46,47]

How much sequencing depth is needed? Determining the number of reads required is a tradeoff between the minimal numbers of reads needed to allow an informative and statistical significant analysis and the available budget. This choice is driven by both the platforms and your experimental needs such as the previous knowledge of the relative abundance of your organism(s) or metabolic pathway(s) of interest. If your metagenome or metatranscriptome is of a relatively low complexity, you can use available genome sequences to evaluate the coverage needed.[48] For a metatranscriptome, you'll have to adapt the sequencing coverage if your focus is the most abundant transcripts or the rare transcripts. It has been shown that millions of 16S rRNA reads do not appreciably increase the extracted information and that a cost-efficient read number is sufficient to discriminate adjacent sites.[1,9] In contrast, during the analysis of the stool microbiome of 100 individuals, increasing the depth of sequencing from 4.5 to 11.7 Gb on average per sample, the human fecal gene catalog increased from 3.3[49] to 5.1 million nonredundant microbial genes,[8] respectively.

Multiplexing of samples has both decreased the cost and allowed to control the number of reads for batch of samples. This approach tags each sample with a unique barcode that is also sequenced. The post-sequencing computing pipeline allows the reads to be binned based on the sample of origin, allowing many samples to be simultaneously sequenced.[50] Additional hidden costs that should be kept in mind are library construction required for preparing the DNA to be sequenced, kits, consumables, labor, instrument initial costs and maintenance, personnel support, indirect cost rate, etc. Further additional costs might be associated with the bioinformatics required for filtering low-quality reads, sequence assembly for pair-ended reads, removing human origin contaminating reads, providing raw or processed reads to your laboratory, and data submission. It's a discussion that you may want to have upfront with your collaborator and/or your sequencing facility of choice.

Read quality is always a parameter to take into account. One of the most common metrics for assessing sequence quality data is the Q score. Low Q scores (below 20) can lead to increase false-positive variants. Q_{20}, which represents an error probability of 1%, is an accepted community standard for a high-quality base, similar to the expectation of Sanger-based DNA sequencing. As the technologies improve, we can expect quality standard of Q_{30} (error probability of 1–1000) and above to be the norm.

DATA STORAGE AND DATA ANALYSIS

Next-generation sequencing moved us from the kilo- and megabytes size files to the mega- and terabytes size file world. Although this might not be of great importance when you are performing a single metagenomics experiment, it can quickly become an issue in large-scale studies. To put this in context, the HMP 16S rRNA-targeted approach generated about 250 megabases, while the shotgun sequencing approach produced over 3 terabases.[8] While the former can be handled on a traditional computer, the latter requires a lot of computing time (or CPU hours) on a computer or computer cluster with another class of technical specifications. An alternative is the use of remote or cloud computing power through virtual machine approaches.[51] Be sure that the data and related information is secured during transfer and in the cloud.

When focusing on 16S rRNA-targeted approach, the availability of packaged analysis pipelines greatly facilitates the process. Mothur and QIIME are not the only available options, although both have shown consistency of improvement and regular updates over the last few years.[52,53] These pipelines include statistical tools that allow a complete analysis of your dataset including your metadata. As we have been focusing on the quality of the input and output of metagenomic analysis, it is important to note that the denoising step is a crucial step that can increase microbial diversity (up to a meaningless amount if read quality filtering and chimera removal are not performed) or restrict the observed diversity based on the settings.[54] There is a balance that must be attained; however, this can be a bit more difficult to achieve when conducting the investigation of an understudied microbiome.

Whole genome shotgun sequencing leads to the information about the DNA and/or the RNA in the sample. The applications can and have been numerous. The focus might be on metabolism, discovering new metabolic or antibiotic pathways, phylogeny, site comparison, the distribution of single nucleotide polymorphism in the microbiome(s), the influence of cancer or antibiotic treatment, the behavioral effect, etc. From the same dataset, phylogenetic placement of the microbiota present in the sample can be obtained from the gene pool instead of the 16S rRNA gene as their relative abundance in the dataset is low.[8,55,56] Packaged analysis pipelines including statistical tools are available to download or as an Internet resource. An incomplete list of those resources includes CAMERA,[57] EBI metagenomics,[58] IMG/M,[59] MEGAN,[60] METAREP,[61] and MG-RAST.[62] For all metagenomics applications, commercial software replace or complement freely available tools.

All bioinformatics tools rely on databases to add layers of information, from phylogeny to function. While some are based on only one technology (such as the gene catalogs from METAhit and the HMP), others have evolved through generations of approaches and technological advancements such as COG,[63] KEGG,[64] GenBank, and all the other international depositories.[65] The lack of standardization, inconsistent annotation, and the different technologies leading to specific errors unknown to the investigator create some challenges. Curated databases are attempts to limit those issues and often decrease the dataset size by removing information (e.g., sequences) not relevant to the focus in question. Some of these databases include CAZy,[66] Greengenes,[67] HOMD,[14] and MetaCyc.[68] The power of additional layers of information is in their enrichment of the content that we can derive from a dataset. However, we should keep in mind that part of the information from the dataset is unavailable as it did not perfectly match to a previously obtained dataset. With the diversity of microbial strains yet to be sequenced, the answer to your scientific question might reside in the conserved proteins without associated function, or gene(s) or gene set that have never been deposited before.

DATA AND PUBLICATION

Any metagenomic project should include a plan for sharing the data collected to the scientific community, including sequence data and metadata. The International Nucleotide Sequence Database Collaboration

(INSDC, http://insdc.org) hosts some of the repositories for the collection and dissemination of nucleic acid datasets. INSDC is a joint effort hosting the following computerized databases: DNA Data Bank of Japan (Japan), GenBank (USA), and the European Nucleotide Archive (based in the United Kingdom).[69]

The need to archive well-defined contextual metadata has been recognized by the community, leading to the creation of the Genomic Standards Consortium. Their mission is to work toward: 1) the implementation of new genomic standards, 2) methods of capturing and exchanging the information captured in these standards, and 3) harmonization of information collection and analysis efforts across the wider genomics community.[70] From this effort arose the creation of minimum information requirement for both genomes and metadata to be submitted to the journal and sequence repositories. The MIGS (minimal information about a genome sequence), MIMS (minimal information about metagenome sequence), MIMARKS (minimal information about marker gene sequence), and MIxS (minimum Information about any (X) sequence) specifications are checklists that both standardize and enhance our ability to further analyze datasets for either training or complementary analysis.[71,72] The adoption of such standards elevates the quality, accessibility, and utility of the information collected by the data repository.

As of yet, there is no standard format to present how the data was analyzed. In the best interests of all, the format should include the methods, tools, and parameters used in the analysis. One option is to make the information available as an online appendix to the published article. There is no such thing as pressing a button and getting the completed analysis. Professional scientists, students, and citizen scientists encounter the same issues. Similar standards of high quality should be put into service for the benefits of the biosphere.

LET'S TALK ABOUT THE STATUS QUO

In science, the *status quo*, the existing state of affairs, and the dogma, the established opinion and doctrine, often go hand in hand. Every time a new technology challenges, the *status quo* resistance occurs, not always in the most constructive of ways. It is not our place to choose for you where you stand in the debate regarding the progresses supported by metagenomic approaches. One clear progress is the flow of data. It creates more statistical power to discriminate the aspect(s) of your hypothesis validation, and offers opportunities for validating previously published hypothesis and for hypothesis generation.

What about the "old data," the ones published using more restricted or better focused analyses? There is no current methodology that can yet replace quantitative PCR for detecting the relative abundance of host versus microbial genetic abundance. The previous approaches for cultivation-independent analyses are here to guide us by facilitating the analysis and providing the trampoline needed for the next discovery. The high dimensionality of the datasets is potentially a challenge, but it also brings new opportunities to create a validated system biology approach to better understand biological function.

The conceptual and practical details are project specific and all partners should be part of the discussion and project building (primary investigator, co-investigators, statistician, bioinformatician, core facilities, providers, suppliers, IT department, etc.). This is a call to students, professional scientists, and citizen scientists alike, to create new datasets and tools that are needed. Please research, share, and disseminate.

ACKNOWLEDGMENTS

We would like to thank the colleagues who shared their frustrations and successes, and Rebecca S. Misra for drawing Figure 1.1. We apologize to those colleagues whose work was not included for lack of space. The laboratory is supported by grants CA166150 and DK097153 from the National Institutes of Health.

REFERENCES

1. Caporaso JG, Lauber CL, Walters WA, Berg-Lyons D, Lozupone CA, Turnbaugh PJ, et al. Global patterns of 16S rRNA diversity at a depth of millions of sequences per sample. Proc Natl Acad Sci USA 2011;108(Suppl 1):4516–22.
2. Centers for Disease Control and Prevention. CDC Health Disparities and Inequalities Report – United States, 2013. MMWR 2013;62(3).
3. Venter JC, Adams MD, Myers EW, Li PW, Mural RJ, Sutton GG, et al. The sequence of the human genome. Science 2001;291(5507):1304–51.
4. Human Microbiome Project Consortium. Structure, function and diversity of the healthy human microbiome. Nature 2012;486(7402):207–14.

5. McFall-Ngai M, Hadfield MG, Bosch TC, Carey HV, Domazet-Loso T, Douglas AE, et al. Animals in a bacterial world, a new imperative for the life sciences. Proc Natl Acad Sci USA 2013;110(9):3229–36.

6. Segata N, Haake SK, Mannon P, Lemon KP, Waldron L, Gevers D, et al. Composition of the adult digestive tract bacterial microbiome based on seven mouth surfaces, tonsils, throat and stool samples. Genome Biol 2012;13(6):R42.

7. Giannoukos G, Ciulla D, Huang K, Haas B, Izard J, Levin J, et al. Efficient and robust RNA-seq process for cultured bacteria and complex community transcriptomes. Genome Biol 2012;13(3):R23.

8. Human Microbiome Project Consortium. A framework for human microbiome research. Nature 2012;486(7402):215–21.

9. La Rosa PS, Brooks JP, Deych E, Boone EL, Edwards DJ, Wang Q, et al. Hypothesis testing and power calculations for taxonomic-based human microbiome data. PloS One 2012;7(12):e52078.

10. Tickle TL, Segata N, Waldron L, Weingart U, Huttenhower C. Two-stage microbial community experimental design. ISME J 2013;7(12):2330–9.

11. Lander ES, Waterman MS. Genomic mapping by fingerprinting random clones: a mathematical analysis. Genomics 1988;2(3):231–9.

12. Wendl MC, Kota K, Weinstock GM, Mitreva M. Coverage theories for metagenomic DNA sequencing based on a generalization of Stevens' theorem. J Math Biol 2012;67(5):1141–1161.

13. Pagani I, Liolios K, Jansson J, Chen IM, Smirnova T, Nosrat B, et al. The Genomes On-Line Database (GOLD) v.4: status of genomic and metagenomic projects and their associated metadata. Nucleic Acids Res 2012;40(Database issue):D571–9.

14. Chen T, Yu WH, Izard J, Baranova OV, Lakshmanan A, Dewhirst FE. The Human Oral Microbiome Database: a web accessible resource for investigating oral microbe taxonomic and genomic information. Database: J Biol Databases Curation 2010:baq013.

15. Li X, Du D. Variation, evolution, and correlation analysis of C+G content and genome or chromosome size in different kingdoms and phyla. PloS One 2014;9(2):e88339.

16. Sabath N, Ferrada E, Barve A, Wagner A. Growth temperature and genome size in bacteria are negatively correlated, suggesting genomic streamlining during thermal adaptation. Genome Biol Evol 2013;5(5):966–77.

17. Costello EK, Lauber CL, Hamady M, Fierer N, Gordon JI, Knight R. Bacterial community variation in human body habitats across space and time. Science 2009;326(5960):1694–7.

18. Gerber GK, Onderdonk AB, Bry L. Inferring dynamic signatures of microbes in complex host ecosystems. PLoS Comput Biol 2012;8(8):e1002624.

19. Segata N, Haake SK, Mannon P, Lemon KP, Waldron L, Gevers D, et al. Composition of the adult digestive tract bacterial microbiome based on seven mouth surfaces, tonsils, throat and stool samples. Genome Biol 2012;13(6):R42.

20. Grice EA, Kong HH, Renaud G, Young AC, Bouffard GG, Blakesley RW, et al. A diversity profile of the human skin microbiota. Genome Res 2008;18(7):1043–50.

21. Zhang Z, Geng J, Tang X, Fan H, Xu J, Wen X, et al. Spatial heterogeneity and co-occurrence patterns of human mucosal-associated intestinal microbiota. ISME J 2013;(4):881–893.

22. Gonzalez A, Stombaugh J, Lauber CL, Fierer N, Knight R. SitePainter: a tool for exploring biogeographical patterns. Bioinformatics 2011;28(3):436–438.

23. Slaughter JC, Lumley T, Sheppard L, Koenig JQ, Shapiro GG. Effects of ambient air pollution on symptom severity and medication use in children with asthma. Ann Allergy, Asthma Immunol 2003;91(4):346–53.

24. Goleva E, Jackson LP, Harris JK, Robertson CE, Sutherland ER, Hall CF, et al. The effects of airway microbiome on corticosteroid responsiveness in asthma. Am J Res Crit Care Med 2013;188(10):1193–201.

25. Gilbert JA, Field D, Swift P, Thomas S, Cummings D, Temperton B, et al. The taxonomic and functional diversity of microbes at a temperate coastal site: a "multi-omic" study of seasonal and diel temporal variation. PloS One 2010;5(11):e15545.

26. Dewhirst FE, Chen T, Izard J, Paster BJ, Tanner AC, Yu WH, et al. The human oral microbiome. J Bacteriol 2010;192(19):5002–17.

27. Zeeuwen PL, Boekhorst J, van den Bogaard EH, de Koning HD, van de Kerkhof PM, Saulnier DM, et al. Microbiome dynamics of human epidermis following skin barrier disruption. Genome Biol 2012;13(11):R101.

28. Luo C, Tsementzi D, Kyrpides N, Read T, Konstantinidis KT. Direct comparisons of Illumina vs. Roche 454 sequencing technologies on the same microbial community DNA sample. PloS One 2012;7(2):e30087.

29. Aird L, Anderson S, Jumpstart Consortium Human Microbiome Project Data Generation Working Group. et al. Evaluation of 16S rDNA-based community profiling for human microbiome research. PloS One 2012;7(6):e39315.

30. Franzosa EA, Morgan XC, Segata N, Waldron L, Reyes J, Earl AM, et al. Relating the metatranscriptome and metagenome of the human gut. Proc Natl Acad Sci USA 2014; 111(22):E2329–2338.

31. Dhaliwal A. DNA extraction and purification. Mater Methods 2013;3:191.

32. Watson RJ, Blackwell B. Purification and characterization of a common soil component which inhibits the polymerase chain reaction. Can J Microbiol 2000;46(7):633–42.

33. Yoshikawa H, Dogruman-Al F, Turk S, Kustimur S, Balaban N, Sultan N. Evaluation of DNA extraction kits for molecular diagnosis of human blastocystis subtypes from fecal samples. Parasitol Res 2011;109(4):1045–50.

34. John BA, Okello JZ, Devault Alison M, Kuch Melanie, Okwi Andrew L, Nelson K, et al. Comparison of methods in the recovery of nucleic acids from archival formalin-fixed paraffin-embedded autopsy tissues. Anal Biochem 2010;400:110–7.

35. Su JM, Perlaky L, Li XN, Leung HC, Antalffy B, Armstrong D, et al. Comparison of ethanol versus formalin fixation on preservation of histology and RNA in laser capture microdissected brain tissues. Brain Pathol 2004;14(2):175–82.

36. Chassy BM, Giuffrida A. Method for the lysis of Gram-positive, asporogenous bacteria with lysozyme. Appl Environ Microbiol 1980;39(1):153–8.

37. Frostegard A, Courtois S, Ramisse V, Clerc S, Bernillon D, Le Gall F, et al. Quantification of bias related to the extraction of DNA directly from soils. Appl Environ Microbiol 1999;65(12):5409–20.

38. Salonen A, Nikkila J, Jalanka-Tuovinen J, Immonen O, Rajilic-Stojanovic M, Kekkonen RA, et al. Comparative analysis of fecal DNA extraction methods with phylogenetic microarray: effective recovery of bacterial and archaeal DNA using mechanical cell lysis. J Microbiol Methods 2010;81(2):127–34.

39. Morgan JL, Darling AE, Eisen JA. Metagenomic sequencing of an in vitro-simulated microbial community. PloS One 2010;5(4):e10209.

40. Nylund L, Heilig HG, Salminen S, de Vos WM, Satokari R. Semi-automated extraction of microbial DNA from feces for qPCR and phylogenetic microarray analysis. J Microbiol Methods 2010;83(2):231–5.

41. Feehery GR, Yigit E, Oyola SO, Langhorst BW, Schmidt VT, Stewart FJ, et al. A method for selectively enriching microbial DNA from contaminating vertebrate host DNA. PloS One 2013;8(10):e76096.

42. Wang HX, Geng ZL, Zeng Y, Shen YM. Enriching plant microbiota for a metagenomic library construction. Environ Microbiol 2008;10(10):2684–91.

43. Okubara P, Li C, Schroeder K, Schumacher R, Lawrence N. Improved extraction of *Rhizoctonia* and *Pythium* DNA from wheat roots and soil samples using pressure cycling technology. Can J Plant Pathol 2007;29(3):304–10.

44. Zimmerman N, Izard J, Klatt C, Zhou J, Aronson E. The unseen world: environmental microbial sequencing and identification methods for ecologists. Front Eco Environ 2014;12(4):224–31.

45. Institute NHGR. DNA Sequencing Costs: Data from the NHGRI Genome Sequencing Program 2014 (Accessed February 12, 2014). http://www.genome.gov/sequencingcosts/

46. Dark MJ. Whole-genome sequencing in bacteriology: state of the art. Infect Drug Resist. 2013;6:115–23.

47. Quail MA, Smith M, Coupland P, Otto TD, Harris SR, Connor TR, et al. A tale of three next generation sequencing platforms: comparison of Ion Torrent, Pacific Biosciences and Illumina MiSeq sequencers. BMC Genomics 2012;13:341.

48. Haas BJ, Chin M, Nusbaum C, Birren BW, Livny J. How deep is deep enough for RNA-Seq profiling of bacterial transcriptomes? BMC Genomics 2012;13:734.

49. Qin J, Li R, Raes J, Arumugam M, Burgdorf KS, Manichanh C, et al. A human gut microbial gene catalogue established by metagenomic sequencing. Nature 2010;464(7285):59–65.

50. Andersson AF, Lindberg M, Jakobsson H, Backhed F, Nyren P, Engstrand L. Comparative analysis of human gut microbiota by barcoded pyrosequencing. PloS One 2008;3(7):e2836.

51. Angiuoli SV, Matalka M, Gussman A, Galens K, Vangala M, Riley DR, et al. CloVR: a virtual machine for automated and portable sequence analysis from the desktop using cloud computing. BMC Bioinform 2011;12:356.

52. Schloss PD, Westcott SL, Ryabin T, Hall JR, Hartmann M, Hollister EB, et al. Introducing mothur: open-source, platform-independent, community-supported software for describing and comparing microbial communities. Appl Environ Microbiol 2009;75(23):7537–41.

53. Caporaso JG, Kuczynski J, Stombaugh J, Bittinger K, Bushman FD, Costello EK, et al. QIIME allows analysis of high-throughput community sequencing data. Nat Methods 2010;7(5):335–6.

54. Gaspar JM, Thomas WK. Assessing the consequences of denoising marker-based metagenomic data. PloS One 2013;8(3):e60458.

55. Segata N, Waldron L, Ballarini A, Narasimhan V, Jousson O, Huttenhower C. Metagenomic microbial community profiling using unique clade-specific marker genes. Nat Methods 2012;9(8):811–4.

56. Tu Q, He Z, Zhou J. Strain/species identification in metagenomes using genome-specific markers. Nucleic Acids Res 2014;42(8):e67.

57. Sun S, Chen J, Li W, Altintas I, Lin A, Peltier S, et al. Community cyberinfrastructure for advanced microbial ecology research and analysis: the CAMERA resource. Nucleic Acids Res 2011;39(Database issue):D546–51.

58. Hunter S, Corbett M, Denise H, Fraser M, Gonzalez-Beltran A, Hunter C, et al. EBI metagenomics: a new resource for the analysis and archiving of metagenomic data. Nucleic Acids Res 2014;42(Database issue):D600–6.

59. Markowitz VM, Ivanova NN, Szeto E, Palaniappan K, Chu K, Dalevi D, et al. IMG/M: a data management and analysis system for metagenomes. Nucleic Acids Res 2008;36(Database issue):D534–8.

60. Huson DH, Mitra S, Ruscheweyh HJ, Weber N, Schuster SC. Integrative analysis of environmental sequences using MEGAN4. Genome Res 2011;21(9):1552–60.

61. Goll J, Rusch DB, Tanenbaum DM, Thiagarajan M, Li K, Methe BA, et al. METAREP: JCVI metagenomics reports – an open source tool for high-performance comparative metagenomics. Bioinformatics 2010;26(20):2631–2.

62. Meyer F, Paarmann D, D'Souza M, Olson R, Glass EM, Kubal M, et al. The metagenomics RAST server – a public resource for the automatic phylogenetic and functional analysis of metagenomes. BMC Bioinform 2008;9:386.

63. Tatusov RL, Galperin MY, Natale DA, Koonin EV. The COG database: a tool for genome-scale analysis of protein functions and evolution. Nucleic Acids Res 2000;28(1):33–6.

64. Kanehisa M, Goto S, Furumichi M, Tanabe M, Hirakawa M. KEGG for representation and analysis of molecular networks involving diseases and drugs. Nucleic Acids Res 2010;38 (Database issue):D355–60.

65. Benson DA, Karsch-Mizrachi I, Lipman DJ, Ostell J, Wheeler DL. GenBank. Nucleic Acids Res 2008;36(Database issue):D25–30.

66. Lombard V, Golaconda Ramulu H, Drula E, Coutinho PM, Henrissat B. The carbohydrate-active enzymes database (CAZy) in 2013. Nucleic Acids Res 2014;42(1):D490–5.

67. DeSantis TZ, Hugenholtz P, Larsen N, Rojas M, Brodie EL, Keller K, et al. Greengenes, a chimera-checked 16S rRNA gene database and workbench compatible with ARB. Appl Environ Microbiol 2006;72(7):5069–72.

68. Caspi R, Altman T, Dale JM, Dreher K, Fulcher CA, Gilham F, et al. The MetaCyc database of metabolic pathways and enzymes and the BioCyc collection of pathway/genome databases. Nucleic Acids Res 2010;38(Database issue):D473–9.

69. Cochrane G, Karsch-Mizrachi I, Nakamura Y. The International Nucleotide Sequence Database Collaboration. Nucleic Acids Res 2011;39(Database issue):D15–8.

70. Field D, Amaral-Zettler L, Cochrane G, Cole JR, Dawyndt P, Garrity GM, et al. The Genomic Standards Consortium. PLoS Biol 2011;9(6):e1001088.

71. Yilmaz P, Kottmann R, Field D, Knight R, Cole JR, Amaral-Zettler L, et al. Minimum information about a marker gene sequence (MIMARKS) and minimum information about any (x) sequence (MIxS) specifications. Nat Biotechnol 2011;29(5):415–20.

72. Field D, Garrity G, Gray T, Morrison N, Selengut J, Sterk P, et al. The minimum information about a genome sequence (MIGS) specification. Nat Biotechnol 2008;26(5):541–7.

CHAPTER 2

Long-Read, Single Molecule, Real-Time (SMRT) DNA Sequencing for Metagenomic Applications

Brett Bowman, Mincheol Kim, Yong-Joon Cho and Jonas Korlach

Elucidating the Earth's microecology remains one of the foremost challenges in biology, with profound implications for human health, agriculture, chemistry, energy, and other areas. We have thus far only captured a very small fraction of the Earth's microbial diversity, with estimates of the number of bacterial and archaeal "species" reaching into the millions.[1] However, our understanding of microbial communities has been dramatically improving through the use of high-throughput DNA sequencing technologies.

The sequencing of ribosomal RNA (rRNA) genes, in particular, the small subunits (SSUs), have been widely used for over 30 years for studying microbial community structure, despite limitations imposed by DNA sequencing technologies.[2] For years, the only method available was to painstakingly clone each individual gene of interest, tile over it with multiple Sanger sequencing reactions, and manually stitch the results together.[3] As recently as 2008, Sanger sequencing was still the most common approach, as contemporary next-generation sequencers with read lengths of 100 base pairs or less were unable to significantly differentiate taxa.[4]

This changed rapidly starting around 2009 with the introduction of the titanium sequencing chemistry for 454 pyrosequencing, providing read lengths of greater than 300 bases for hundreds of thousands of reads at a time.[5] Simultaneously, the development of specialized software tools such as Mothur,[6] the RDP classifier,[7] and QIIME[8] allowed the analysis of such large datasets. More recently, this trend has continued with the adoption of approximately 200 base pair assembled paired-end Illumina reads for some metagenomics applications,[9] allowing for sequencing millions of reads in a single experiment, albeit at the cost of reduced read lengths compared with other sequencing technologies. The adoption of next-generation sequencing for metagenomics thus led to an exponential

Metagenomics for Microbiology. http://dx.doi.org/10.1016/B978-0-12-410472-3.00002-6

increase in the amount of data that could be generated from uncultured samples, providing the foundational method for projects such as Metagenomics and Microbial Ecology[10] and the Human Microbiome Project.[11]

However, efforts to obtain clear pictures of metagenomes in this fashion have been complicated by the short read lengths that limit the resolving power of rDNA sequences, as well as inherent biases from both the polymerase chain reaction (PCR) and the next-generation sequencing technologies.[12] The largest source of bias in community 16S sequencing is caused by the initial PCR step.[13] Careful primer selection is important both because different variable regions of the 16S gene show differing capacities to differentiate taxa[14] and no primer sites in the gene are perfectly conserved across all phyla.[15] Read lengths between 500 and 700 bp are sufficient to differentiate most phyla,[16,17] but which regions are required vary, and no region has the resolution of the full-length gene.

In addition, biases inherent in the next-generation sequencing technologies can affect the data interpretation.[18] For 454 sequencing, this led to the development of PyroNoise that attempts to reduce the effect of context-specific error on the analysis of amplicons.[19] To our knowledge, no similar tools have been developed for Illumina-based sequence data, despite the platform also having known context-specific errors.[20] GC-content bias can also affect the quality of the second-generation sequencing data,[21] and this effect has been directly studied on the 454 platform for 16S sequencing.[22,23]

Here, we describe the application of long sequence reads provided by Pacific Biosciences' single molecule, real-time (SMRT) DNA sequencing to decode the entire 16S rRNA gene. SMRT sequencing is based on monitoring the activity of individual DNA polymerase molecules and detecting its activity of successive nucleotide incorporations in real time.[24,25]. Compared with other sequencing methods, it exhibits much longer read lengths (8500 bp on average with the latest (P5-C3) sequencing chemistry), the least sequence context bias,[26] and a high consensus accuracy due to the random nature of sequencing errors.[27] By exploiting these characteristics and moving from shorter amplicons to sequencing the full-length gene significantly reduces the primer bias in 16S community profiling. The selection of specific variable regions, important when read length is a limiting factor, is no longer required as the entire sequence is obtained in a single read. Although biases inherent in primer designs are unavoidable, the sites flanking the terminal V1 and V9 regions are among the most conserved: primers targeting those sites, commonly referred to as either 27F/1492R or GM3/GM4, are among the most extensively used and optimized 16S primers,[28] capturing approximately 87% of known sequences with less than two mismatches.[15] In addition, following the initial PCR, there are no additional amplification steps during library preparation and sequencing, avoiding any further amplification bias. SMRT sequencing has been shown to display very little bias with respect to GC content and sequence context,[29] resulting in higher sequence quality across the entire 16S rRNA gene and reduced bias in community structure.

FULL-LENGTH 16S rRNA GENE SEQUENCING

To demonstrate the application of SMRT sequencing to surveying metagenomic amplicons, we sequenced a metagenomic mock community consisting of an equimolar mixture of 20 known, full-length 16S rRNA gene sequences from 12 distinct bacterial lineages. We analyzed PCR-amplified, full-length 16S rRNA genes using 27F/1492R primers and prepared sequencing libraries from the amplicons according to the standard library preparation protocol.[30] Sequencing was performed in triplicate by running three barcoded technical PCR replicates on each SMRT Cell. To generate high-quality, full-length 16S sequence reads, we employed circular consensus sequencing (CCS),[30] which allows for the repeated sequencing of the same DNA molecule to generate a high-quality intramolecular consensus (Figure 2.1A). The median read length for sequenced molecules was 5560 bp or approximately 3.5 passes over the ~1500 bp template sequence. Each SMRT Cell produced 31,000–43,000 raw sequence reads, of which 17,000–24,000 reads contained sufficient coverage of the template to generate CCS sequences. It is worth noting that the samples were somewhat underloaded suggesting that even greater throughput could be achieved upon optimizing loading conditions. Comparison of the predicted CCS read accuracy with the known reference sequences showed excellent concordance, as calculated from the per-base phred quality scores (Figure 2.1B), with a median predicted accuracy of 99.7% over all reads (Figure 2.1C).

The sequences were analyzed with a combination of standard tools available in Mothur[6] and custom python scripts to accommodate the unique needs of single-molecule sequencing data, collectively available

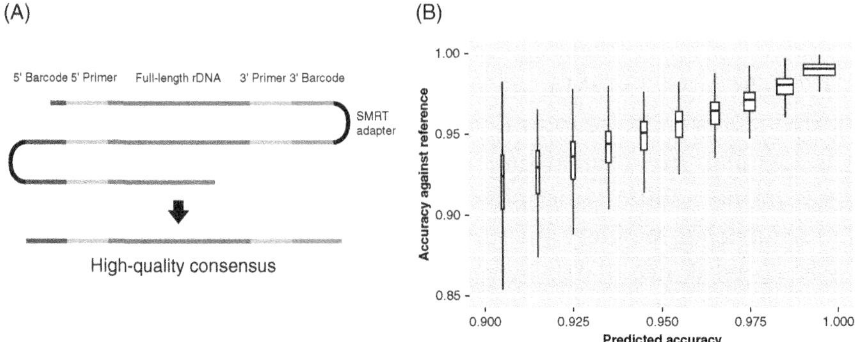

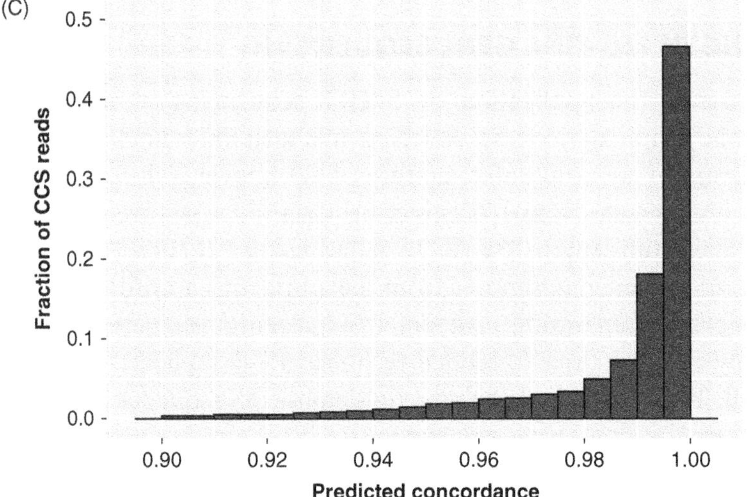

Fig. 2.1. SMRT Sequencing of full-length 16S RNA generated from a mock community of 20 known sequences. (A) Schematic of generating high-accuracy 16S reads through circular consensus sequencing (CCS). (B) Concordance of predicted CCS accuracy versus observed accuracy against the mock community reference. (C) Histogram of predicted concordance with the reference for full-length 16S CCS sequences.

for public use on Github as rDnaTools.[31] Sequences from different replicates were demultiplexed if at least one barcode sequence could be identified with HMMER,[32] which recovered 99.5% of all CCS sequences. Truncated sequences under 500 bp and concatenated products over 2000 bp were discarded. De-multiplexed sequences were then aligned to the SILVA reference alignment of bacterial ribosomal SSU sequences. Despite a range of sequence lengths (1483 ± 169 bp), 98.5% of all de-multiplexed sequences covered the entire canonical alignment (Figure 2.2). The differences in lengths are because of the biological

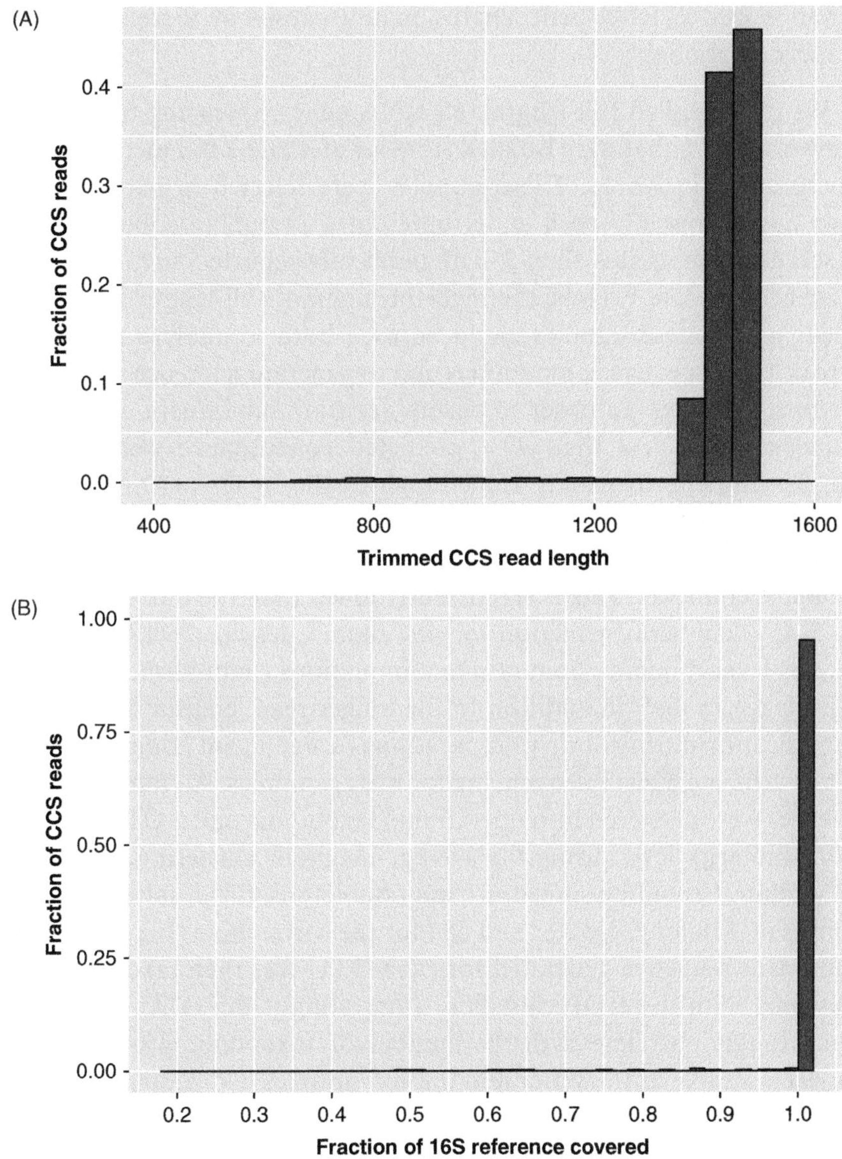

Fig. 2.2. Sequence lengths of 16S rRNA gene SMRT sequencing CCS reads from a mock community of 20 known sequences. (A) Sequence lengths after barcode trimming and chimera filtering. (B) Fraction of 16S rRNA genes covered by the sequences after SILVA alignment.

variation of 16S RNA gene lengths, mainly caused by more variability in the loop regions.

We then applied full-length 16S RNA gene sequencing to two unknown samples that were isolated from a water and a soil metagenomic environment in Korea. These samples were selected in part because Asia has historically been underrepresented in public sequence databases, comprising less than 3% of publically reported environmental sequences to date.[33] Thus, the high-throughput, full-length rDNA sequencing of these samples could be used both to analyze the structure of the respective communities and to generate reference sequences for future studies. In order to ensure accurate subsequent clustering, sequences with less than 99% predicted concordance were filtered out, leaving approximately 15,000 high-quality reads per SMRT Cell for analysis. Chimera detection was carried out with the Mothur implementation of Uchime,[34] and 2.4% of sequences were removed as probable chimeras. This is significantly lower than the published rates of 5–45% chimera formation for 454 data,[35] despite >99% recall of species-level chimeras from *in silico* simulations (unpublished results). This suggests that, in addition to the longer read lengths, the lack of amplification during library preparation is a significant additional benefit to utilizing SMRT sequencing for metagenomics. All remaining sequences were clustered into operational taxonomic units (OTUs) at the 97% similarity level, using the average neighbor clustering algorithm in Mothur. Excluding singletons and doubletons, this resulted in OTU counts of 318 for the water and 684 for the soil sample (Figure 2.3). A consensus sequence generated for each OTU was then taxonomically classified with the RDP classifier.[7] Interestingly, the OTU counts do not strongly correlate with the number of taxonomic groups found for each sample as the water and the soil samples had similar numbers of taxa (137 for water vs. 100 for soil, Figure 2.3), but a greater than twofold difference in OTUs (318 for water vs. 684 for soil, Figure 2.3). This difference highlights the increase in resolution power provided by full-length 16S sequences. Short reads are typically limited in taxa resolution, whereas the full-length sequences provide insights into the level and distribution within taxa that provide the means to expand the information depth of metagenomic databases and develop tools to characterize this diversity.

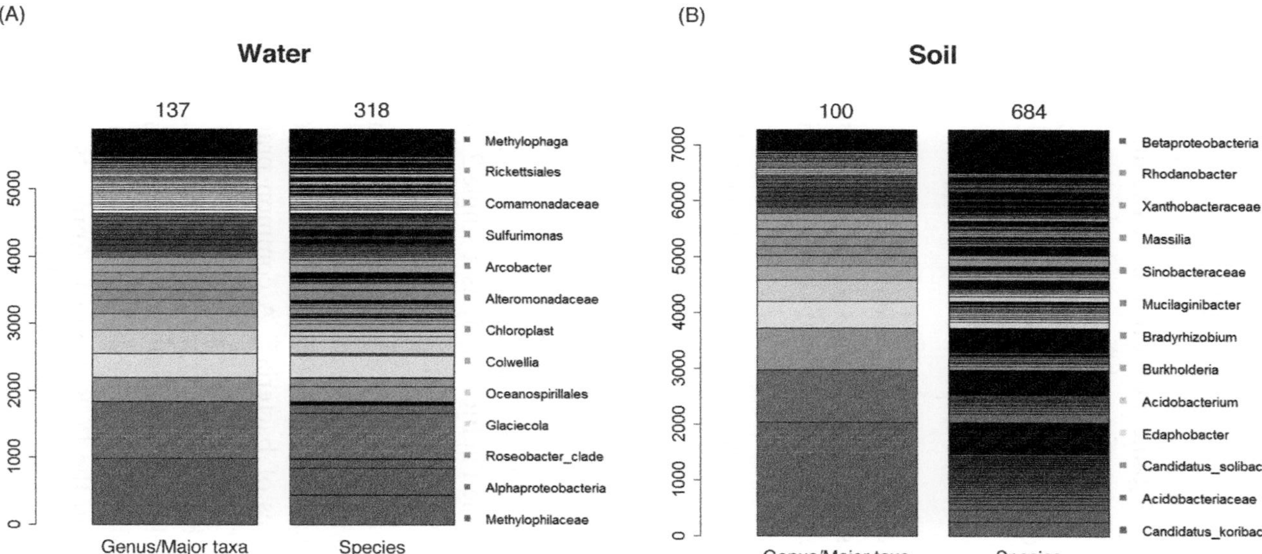

Fig. 2.3. Composition of environmental (A) aquatic and (B) soil samples as determined by 16S rRNA gene sequencing using full-length SMRT Sequencing. Left panel: taxonomic assignments; right panel: OTUs. The numbers on the top of the panels indicate the total frequencies obtained from one SMRT Cell. The legend on the right only contains a partial list of taxa for illustration purposes and is not meant to be exhaustive. Regions of black in the composition graphs are because of low abundance groups that get fused by the black line thickness.

The consensus sequences of the largest OTUs from each identified taxa were used to construct a phylogenetic tree for each community. The abundances of CCS reads and OTUs were tabulated for each taxonomic group and plotted next to their associated leaf on the tree (Figure 2.4). It is worth noting that while the two distributions are highly similar in the soil sample, they significantly diverge in the aquatic community. These differences may stem from differential selective pressures between OTUs, which would be more difficult to quantify without the resolution of clustering enabled by full-length rDNA sequencing.

DISCUSSION

Although metagenomic community identification has become the primary use for DNA sequencing, many service providers world-wide still use 16S rRNA gene to identify tens of thousands of individual bacterial isolates, relying on the expensive and labor-intensive approach of cloning and tiling Sanger reads over each sequence[16]. The application of high-throughput, multiplexed, full-length 16S SMRT sequencing[36] demonstrates an alternative approach to strain identification. By sequencing the whole gene in a single reaction, it becomes possible to avoid cloning and obtain the full gene sequence directly from the PCR product. In addition, the ability to sequence multiple amplicons, multiplexing, onto a single SMRT Cell offers the possibility of replacing a bank of capillary sequencers with a single PacBio RS II, as a single 2-hour run can provide complete sequences for dozens of amplicons.

Another limitation of current approaches to strain identification is the limited resolution provided by the 16S rRNA gene alone[37]. It is well known that even the full-length 16S gene is insufficient to accurately separate some lineages of bacteria[38]. A common practice has been to perform restriction fragment-length polymorphism (RFLP) of the intergenic spacer region (ISR) between genes in the ribosomal operon instead of sequencing homologous populations[39]. However, this method is only cost-effective when the family is already known and the analysis can be carried out directly on the PCR product. Application of this method to unknown or environmental samples additionally requires the separate cloning and sequencing of the rRNA operon for the determination of genus[40]. Since the additional information is not always needed, most

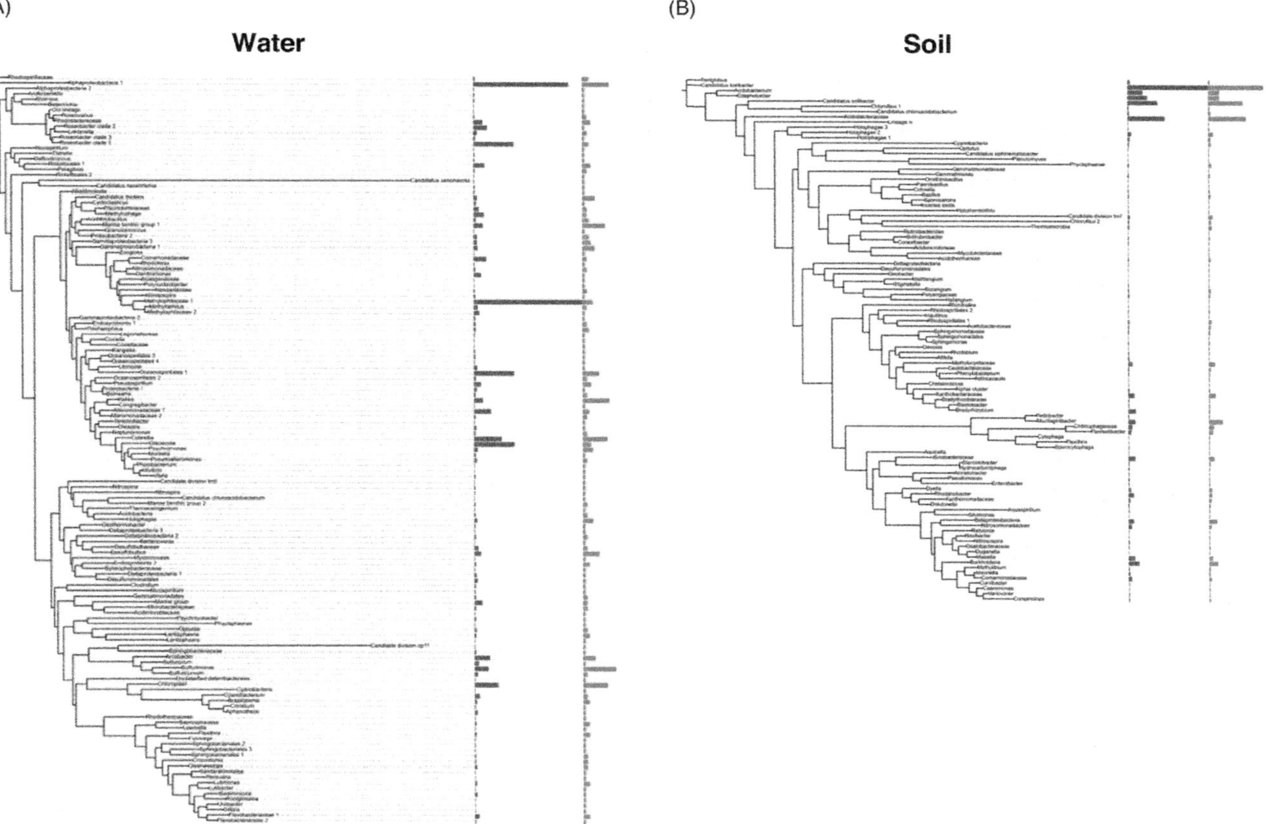

Fig. 2.4. Phylogeny of environmental 16S rRNA samples from (A) an aquatic and (B) a soil metagenomic community. Bars denote the number of CCS sequences (red) and OTUs (green) associated with each taxonomic group.

strain identification has been done with a single amplicon of between 400 and 1,500 bp depending on the resolution required[41].

Ideally, one would perform only a single amplification for the entire *rrn* operon, encompassing at a minimum the 16S and 23S genes in addition to the internal transcribed spacers (ITS)[42]. The combination of multiple, co-linear rRNA genes provides equivalent resolution to multi-locus sequence typing (MLST), while negating the need for multiple PCRs or the generation of new reference databases[43]. This is impractical if the amplicon needs to be cloned in order to be sequenced in parts. A tantalizing possibility is the sequencing of the entire 5 kb length of the operon in a single read. With the full sequence, genus-level identification could be obtained and validated by comparing the classification of the 16S and 23S rRNA genes, while species-level specificity could be obtained from the ITS. The individual component sequences could also be used to fill numerous gaps in current rRNA databases, many of which contain strains not yet sequenced at the other loci[44]. With >15 kb read lengths attainable with SMRT sequencing, the opportunity exists to take this approach, but to our knowledge no groups have done so yet. In general, other markers that had not been adopted due to previous read length limitations may also now be more accessible to sequence analysis, e.g. rpoB[45]. This also applies to 18S rRNA characterization for eukaryotic metagenomes and ITS sequencing for fungal metagenomes.

Another consequence of requiring cloning for strain identification with current protocols is that the resulting sequence is representative of only a single rRNA gene, while it is known that most bacteria have multiple, varying copies of their ribosomal genes[46]. Since intra-species diversity of 16S rRNA genes is small enough that it is unlikely to affect sequence classification[47], this diversity from multiplicity has not been well characterized and is not well captured for most lineages by current rRNA databases. The rRNA operon copy number appears to significantly correlate with the species' ecological strategy, which is of obvious interest in metagenomic studies[48]. If the number of operons can be easily estimated for each lineage, it may be possible to incorporate this information into new methods of measuring metagenomes. Such methods could offer additional insight into the structure and function of microbial communities, beyond the measures of diversity and species richness currently available.

Though ribosomal rRNA gene sequencing has been a foundational tool for metagenomics, it has long been understood that it provides only a limited view of microbial communities. Specifically, outside the context of their originating genomes, rRNA genes provide no direct information about the function of the microbe in its community[49]. Without knowledge of function, it is impossible to determine the causes of a microbe's effect on its environment or the role it plays in the community. This is especially important to understanding the stability of metagenomic communities given the strong evidence for functional redundancy among healthy communities[50]. Methods that have been developed to elucidate these questions have been termed "functional metagenomics"[51].

The first methods for functional metagenomics involved sequencing BACs and fosmids of cloned metagenomic sequence[52], and these methods have already been applied in conjunction with SMRT sequencing, e.g. for the characterization of antibiotic resistance genes in cow manure[53]. The paper highlights the power of the long sequence reads to capture the genomic context of antibiotic resistance genes to infer the taxonomic affiliation of the origin of the DNA which is critical for drawing conclusions about horizontal transfer of antibiotic resistance genes in bacterial populations. With the advent of next-generation sequencing the field has moved towards *de novo* shotgun sequencing of metagenomes for functional analysis[54]. This has greatly increased the throughput with which functional metagenomic data can be generated, but the resulting contigs have typically been more than an order of magnitude smaller than those possible from cloned libraries[55]. The smaller contig sizes pose a challenge for the interpretation of metagenomic sequence data, and multiple methods have been developed that attempt to bin contigs from similar organisms based on their k-mer frequencies. These methods have been limited by the wide diversity of k-mers and GC content within any one genome, coupled with short read lengths and context-specific biases of second-generation sequencing. In principle, long-read sequencing data should enable more detailed functional examinations of metagenomes, requiring less binning due to larger contigs and exhibiting less context-specific bias. Combined with recent advances in generating finished bacterial genomes[56], these capabilities could also improve the characterization of single-cell isolation derived metagenomes[57]. With the ability to identify both species and community structure from metagenomic rDNA, as well as the potential for improving functional metagenomics, we anticipate that SMRT sequencing will open new opportunities for the analysis of metagenomic communities.

REFERENCES

1. Curtis TP, Sloan WT, Scannell JW. Estimating prokaryotic diversity and its limits. Proc Natl Acad Sci USA 2002;99:10494–9.

2. Woese CR, Fox GE. Phylogenetic structure of the prokaryotic domain: the primary kingdoms. Proc Natl Acad Sci USA 1977;74:5088–90.

3. Weisburg WG, Barns SM, Pelletier DA, Lane DJ. 16S ribosomal DNA amplification for phylogenetic study. J Bacteriol 1991;173:697–703.

4. Kunin V, Copeland A, Lapidus A, Mavromatis K, Hugenholtz P. A bioinformatician's guide to metagenomics. Microbiol Mol Biol Rev 2008;72:557–78.

5. Armougom F, Raoult D. Exploring microbial diversity using 16S rRNA high-throughput methods. J Comput Sci Syst Biol 2009;2:74–92.

6. Schloss PD, Westcott SL, Ryabin T, Hall JR, Hartmann M, Hollister EB, et al. Introducing Mothur: open-source, platform-independent, community-supported software for describing and comparing microbial communities. Appl Environ Microbiol 2009;75:7537–41.

7. Wang Q, Garrity GM, Tiedje JM, Cole JR. Naive Bayesian classifier for rapid assignment of rRNA sequences into the new bacterial taxonomy. Appl Environ Microbiol 2007;73:5261–7.

8. Caporaso JG, Kuczynski J, Stombaugh J, Bittinger K, Bushman FD, Costello EK, et al. QIIME allows analysis of high- throughput community sequencing data Intensity normalization improves color calling in SOLiD sequencing. Nat Publ Gr 2010;7:335–6.

9. Bartram AK, Lynch MDJ, Stearns JC, Moreno-Hagelsieb G, Neufeld JD. Generation of multimillion-sequence 16S rRNA gene libraries from complex microbial communities by assembling paired-end illumina reads. Appl Environ Microbiol 2011;77:3846–52.

10. Xu J. Microbial ecology in the age of genomics and metagenomics: concepts, tools, and recent advances. Mol Ecol 2006;15:1713–31.

11. Turnbaugh PJ, Ley RE, Hamady M, Fraser-Liggett CM, Knight R, Gordon JI, et al. The human microbiome project. Nature 2007;449:804–10.

12. Schloss PD, Gevers D, Westcott SL. Reducing the effects of PCR amplification and sequencing artifacts on 16S rRNA-based studies. PLoS One 2011;6:e27310.

13. Baker GC, Smith JJ, Cowan DA. Review and re-analysis of domain-specific 16S primers. J Microbiol Methods 2003;55:541–55.

14. Kumar PS, Brooker MR, Dowd SE, Camerlengo T. Target region selection is a critical determinant of community fingerprints generated by 16S pyrosequencing. PLoS One 2011;6:e20956.

15. Klindworth A, Pruesse E, Schweer T, Peplies J, Quast C, Horn M, et al. Evaluation of general 16S ribosomal RNA gene PCR primers for classical and next-generation sequencing-based diversity studies. Nucleic Acids Res 2013;41:e1.

16. Clarridge JE. Impact of 16S rRNA gene sequence analysis for identification of bacteria on clinical microbiology and infectious diseases. Clin Microbiol Rev 2004;17:840–62.

17. Paster BJ, Boches SK, Galvin JL, Ericson RE, Lau CN, Levanos VA, et al. Bacterial diversity in human subgingival plaque. J Bacteriol 2001;183:3770–83.

18. Harismendy O, Ng PC, Strausberg RL, Wang X, Stockwell TB, Beeson KY, et al. Evaluation of next generation sequencing platforms for population targeted sequencing studies. Genome Biol 2009;10:R32.

19. Quince C, Lanzén A, Curtis TP, Davenport RJ, Hall N, Head IM, et al. Accurate determination of microbial diversity from 454 pyrosequencing data. Nat Meth 2009;6:639–41.

20. Minoche AE, Dohm JC, Himmelbauer H. Evaluation of genomic high-throughput sequencing data generated on Illumina HiSeq and genome analyzer systems. Genome Biol 2011;12:R112.

21. Davey JW, Hohenlohe PA, Etter PD, Boone JQ, Catchen JM, Blaxter ML. Genome-wide genetic marker discovery and genotyping using next-generation sequencing. Nat Rev Genet 2011;12:499–510.

22. Kauserud H, Kumar S, Brysting AK, Nordén J, Carlsen T. High consistency between replicate 454 pyrosequencing analyses of ectomycorrhizal plant root samples. Mycorrhiza 2012;22: 309–15.

23. Pinto AJ, Raskin L. PCR biases distort bacterial and archaeal community structure in pyrosequencing datasets. PLoS One 2012;7:e43093.

24. Eid J, Fehr A, Gray J, Luong K, Lyle J, Otto G, et al. Real-time DNA sequencing from single polymerase molecules. Science 2009;323:133–8.

25. Korlach J, Bjornson KP, Chaudhuri BP, Cicero RL, Flusberg BA, Gray JJ, et al. Real-time DNA sequencing from single polymerase molecules. Methods Enzymol 2010;472:431–55.

26. Ross MG, Russ C, Costello M, Hollinger A, Lennon NJ, Hegarty R, et al. Characterizing and measuring bias in sequence data. Genome Biol 2013;14:R51.

27. Koren S, Harhay GP, Smith TP, Bono JL, Harhay DM, Mcvey SD, et al. Reducing assembly complexity of microbial genomes with single-molecule sequencing. Genome Biol 2013;14:R101.

28. Abed RMM, Al-Thukair A, de Beer D. Bacterial diversity of a cyanobacterial mat degrading petroleum compounds at elevated salinities and temperatures. FEMS Microbiol Ecol 2006;57:290–301.

29. Carneiro MO, Russ C, Ross MG, Gabriel SB, Nusbaum C, DePristo MA. Pacific biosciences sequencing technology for genotyping and variation discovery in human data. BMC Genomics 2012;13:375.

30. Travers KJ, Chin C-S, Rank DR, Eid JS, Turner SW. A flexible and efficient template format for circular consensus sequencing and SNP detection. Nucleic Acids Res 2010;38:e159.

31. Bowman BN. rDnaTools. Available at <https://github.com/PacificBiosciences/rDnaTools (accessed October 8, 2014).

32. Eddy SR. A new generation of homology search tools based on probabilistic inference. Genome Inform 2009;23:205–11.

33. Pagani I, Liolios K, Jansson J, Chen IMA, Smirnova T, Nosrat B, et al. Genomes Online Database. Available at <http://genomesonline.org/> (accessed October 8, 2014).

34. Edgar RC, Haas BJ, Clemente JC, Quince C, Knight R. UCHIME improves sensitivity and speed of chimera detection. Bioinformatics 2011;27:2194–200.

35. Haas BJ, Gevers D, Earl AM, Feldgarden M, Ward DV, Giannoukos G, et al. Chimeric 16S rRNA sequence formation and detection in Sanger and 454-pyrosequenced PCR amplicons. Genome Res 2011;21:494–504.

36. Bowman B, Shin MY, Lee J, Cho Y, Chun J. Analysis of full-length metagenomic 16S genes by SMRT sequencing. Am Soc Microbiol Gen Meet 2013;75.

37. Petti C. Detection and identification of microorganisms by gene amplification and sequencing. Clin Infect Dis 2007;44:1108–14.

38. Janda JM, Abbott SL. 16S rRNA gene sequencing for bacterial identification in the diagnostic laboratory: pluses, perils, and pitfalls. J Clin Microbiol 2007;45:2761–4.

39. Marsh TL. Terminal restriction fragment length polymorphism (T-RFLP): an emerging method for characterizing diversity among homologous populations of amplification products. Curr Opin Microbiol 1999;2:323–7.

40. Wang M, Ahrné S, Antonsson M, Molin G. T-RFLP combined with principal component analysis and 16S rRNA gene sequencing: an effective strategy for comparison of fecal microbiota in infants of different ages. J Microbiol Methods 2004;59:53–69.

41. Tringe SG, Hugenholtz P. A renaissance for the pioneering 16S rRNA gene. Curr Opin Microbiol 2008;11:442–6.

42. Abd-El-Haleem D, Layton AC, Sayler GS. Long PCR-amplified rDNA for PCR-RFLP- and Rep-PCR-based approaches to recognize closely related microbial species. J Microbiol Methods 2002;49:315–9.

43. Zeng YH, Koblížek M, Li YX, Liu YP, Feng FY, Ji JD, et al. Long PCR-RFLP of 16S-ITS-23S rRNA genes: a high-resolution molecular tool for bacterial genotyping. J Appl Microbiol 2013;114:433–47.

44. Munoz R, Yarza P, Ludwig W, Euzéby J, Amann R, Schleifer KH. Release LTPs104 of the all-species living tree. Syst Appl Microbiol 2011;34:169–70.

45. Case RJ, Boucher Y, Dahllöf I, Holmström C, Doolittle WF, Kjelleberg S. Use of 16S rRNA and rpoB genes as molecular markers for microbial ecology studies. Appl Environ Microbiol 2007;73:278–88.

46. Klappenbach JA, Saxman PR, Cole JR, Schmidt TM. rrndb: the Ribosomal RNA Operon Copy Number Database. Nucleic Acids Res 2001;29:181–4.

47. Coenye T, Vandamme P. Intragenomic heterogeneity between multiple 16S ribosomal RNA operons in sequenced bacterial genomes. FEMS Microbiol Lett 2003;228:45–9.

48. Klappenbach JA, Dunbar JM, Schmidt TM. rRNA operon copy number reflects ecological strategies of bacteria. Appl Environ Microbiol 2000;66:1328–33.

49. Gill SR, Pop M, Deboy RT, Eckburg PB, Turnbaugh PJ, Samuel BS, et al. Metagenomic analysis of the human distal gut microbiome. Science 2011;312:1355–9.

50. Ferrer M, Ruiz A, Lanza F, Haange SB, Oberbach A, Till H, et al. Microbiota from the distal guts of lean and obese adolescents exhibit partial functional redundancy besides clear differences in community structure. Environ Microbiol 2013;15:211–26.

51. Dinsdale E, Edwards RA, Hall D, Angly F, Breitbart M, Brulc JM, et al. Functional metagenomic profiling of nine biomes. Nature 2008;452:629–32.

52. Rondon MR, August PR, Bettermann AD, Brady SF, Grossman TH, Liles MR, et al. Cloning the soil metagenome: a Strategy for Accessing the Genetic and Functional Diversity of Uncultured Microorganisms. Appl Environ Microbiol 2000;66:2541–7.

53. Wichmann F, Udikovic-Kolic N. Diverse antibiotic resistance genes in dairy cow manure. mBio 2014;5:e01017–e1113.

54. Venter JC, Remington K, Heidelberg JF, Halpern AL, Rusch D, Eisen JA, et al. Environmental genome shotgun sequencing of the Sargasso sea. Science 2004;304:66–74.

55. Qin J, Li R, Raes J, Arumugam M, Burgdorf KS, Manichanh C, et al. A human gut microbial gene catalogue established by metagenomic sequencing. Nature 2010;464:59–65.

56. Chin C-S, Alexander DH, Marks P, Klammer AA, Drake J, Heiner C, et al. Nonhybrid, finished microbial genome assemblies from long-read SMRT sequencing data. Nat Methods 2013;10:563–9.

57. Rinke C, Schwientek P, Sczyrba A, Ivanova NN, Anderson IJ, Cheng JF, et al. Insights into the phylogeny and coding potential of microbial dark matter. Nature 2013;499:431–7.

CHAPTER 3

Ribosomal RNA Removal Methods for Microbial Transcriptomics

Shaomei He

INTRODUCTION

Transcriptome is the complete set of transcribed sequences in a cell, including messenger RNA (mRNA), ribosomal RNA (rRNA), transfer RNA (tRNA), and regulatory noncoding RNA. Metatranscriptomics, also referred to as "environmental transcriptomics" or "microbial community transcriptomics," studies the entire collection of transcribed sequences from natural communities.[1] As a powerful tool in the post-genomics era, transcriptomics facilitates the understanding of gene structure and regulation, provides a snapshot of gene expression under specific conditions, and reveals regulation of functions responding to different environments.[1–3]

Microbial transcriptomics is evolving with technological advancement, from constructing cDNA clone libraries to oligonucleotide microarrays, and more recently to next-generation sequencing technology-enabled whole transcriptome shotgun sequencing (RNA-Seq). RNA-Seq overcomes many technical constraints in microarrays, such as limited dynamic range, reference genome requirement, and hybridization efficiency and specificity. Together with the increasing throughput and decreasing cost of ultra-high-throughput sequencing, RNA-Seq has become a more popular choice than microarrays for microbial transcriptomics.

A major technical challenge for RNA-Seq is the low relative abundance of mRNA in total cellular RNA (1–5%),[4] the bulk of which is rRNA and tRNA,[5] particularly rRNA. For example, rRNA sequences predominated the generated cDNA library up to >90% of the entire sequences,[6–9] wasting enormous amounts of sequencing resource. Therefore, researchers often chose to eliminate rRNA before sequencing in order to improve mRNA detection sensitivity. Unlike eukaryotic

Metagenomics for Microbiology. http://dx.doi.org/10.1016/B978-0-12-410472-3.00003-8

mRNAs, which have poly(A) tails[10] and can be selectively synthesized into cDNA using poly(T) primers, prokaryotic mRNAs lack poly(A) tails long enough for this application. This had greatly hindered RNA-Seq studies on prokaryotes.

Over the years, different methods have been developed and applied to eliminate prokaryotic rRNA with varied success. These methods employ unique features of rRNA relative to non-rRNA, such as size, sequence, 5′-phosphate, secondary structure, and abundance, and commercial kits are available for some of these methods (Table 3.1). The rRNA elimination step can be performed prior to, during, or after cDNA synthesis. Compared with the elimination during or after cDNA synthesis, the removals prior to cDNA synthesis generally require higher amounts of input RNA (e.g., microgram quantities), and the manipulation is on raw RNA, therefore having a higher risk of mRNA degradation during sample processing.

So which method should be chosen? Two most important considerations in method selection are efficiency and potential bias. Some limitations and biases are obvious, while some are not well recognized. Therefore, a number of benchmarking and validation studies were conducted to systematically evaluate commonly applied methods, using single-species cultures, simple synthetic communities, and complex nature environmental samples.[6–9,11] Owning to the high-throughput sequencing applied, besides assessing the representation rRNA in total sequences, these studies were able to evaluate rRNA removal fidelity, that is, the ability to preserve relative mRNA transcript abundance after rRNA removal, compared with the original crude extract. In light of these findings, methodologies for rRNA removal are discussed in this review with a focus on their advantages, limitations, and potential biases.

SIZE SELECTION BY GEL ELECTROPHORESIS

Intact 5S, 16S, and 23S rRNAs are easily recognizable as distinct sharp bands on electrophoresis gels after size separation. To isolate mRNA, McGrath *et al.*[12] extracted RNA from regions between the 23S and 16S, and between the 16S and 5S rRNA bands on the agarose gel. They tested this method on *Escherichia coli* and a number of environmental samples, and observed increased recovery of mRNA in resulted cDNA clone library.[12]

Table 3.1 Summary of Ribosomal RNA Removal Methods

Method		rRNA Feature Employed	RNA Input	Commercial Kit
Prior to cDNA synthesis				
Size selection by gel electrophoresis		Size	High	
Denaturing high-performance liquid chromatography		Size and interaction with bead matrix		
5′ phosphate-dependent exonuclease digestion		5′-monophosphate of processed RNA		Epicentre' mRNA-ONLY prokaryotic mRNA isolation kit
RNase H digestion		Antisense sequences to rRNA as primers or probes		
Subtractive hybridization	Commercial generic probes			Ambion's MICROBExpress bacterial mRNA enrichment kit; Invitrogen's RiboMinus transcriptome isolation kit; Epicentre's Ribo-Zero rRNA removal kit
	Organism-specific probes			
	Community-specific probes			
Biased 3′ polyadenylation against rRNA		Complex with ribosome and form secondary structure	Low	Ambion's MessageAmp-II bacteria RNA amplification kit
During cDNA synthesis				
Selective priming with not-so random oligomers	Without amplification	Differential occurrence of hexamers in rRNA relative to non-rRNA	Low	
	With PCR amplification		Low	NuGEN's Ovation prokaryotic RNA-Seq system
After cDNA synthesis				
cDNA library normalization	With duplex specific nuclease digestion	High abundance of rRNA	Low	Evrogen's Trimmer-Direct cDNA normalization kit
	With hydroxyapatite chromatography separation			

The limitations of this method are obvious: fragmented rRNA or pre-rRNA not at the regular sizes of intact mature rRNA cannot be removed, while mRNA that co-migrates as rRNA, because of similarity in size and electrophoretic behavior, can be excluded. In addition, the amount of input RNA needs to be high enough for gel visualization and subsequent mRNA recovery, and it is more challenging to maintain RNase-free in gel, buffer, and electrophoresis equipment than in a single tube.

DENATURING HIGH-PERFORMANCE LIQUID CHROMATOGRAPHY

The physical separation of RNA can also be achieved through RNA chromatography with a nonporous alkylated poly(styrene divinylbenzene) bead matrix as the stationary phase.[13] The separation is largely based on RNA size and strength of interaction with the bead matrix. Selective collections can be performed using eluents of different ionic conditions to wash RNA off the bead matrix. Castro *et al.*[14] adopted denaturing high-performance liquid chromatography for RNA separation and removed 78–92% of rRNA extracted from *Corynebacterium pseudobuberculosis*. This method provides fast separation of bulk rRNA and mRNA (within half an hour). However, a high RNA input (e.g., 20 μm[14]) is needed to recover mRNA. In addition, small RNAs elute faster and can be missed from the recovery. Whether the chromatographic profile-based RNA separation is applicable for other microorganisms or for complex environmental samples remains to be examined.

5′ PHOSPHATE-DEPENDENT EXONUCLEASE DIGESTION

A kit based on specific enzymatic degradation of rRNAs (mRNA-ONLY Prokaryotic mRNA Isolation kit) was developed by Epicentre (Madison, WI). Their Terminator 5′-phosphate-dependent exonuclease is a processive 5′ to 3′ exonuclease, selectively degrading RNA molecules with 5′-monophosphates, but not RNAs with 5′-triphosphate or hydroxyl groups. As mature 5S, 16S, and 23S rRNAs are processed from a single pre-rRNA transcript, they possess 5′-monophosphates and are susceptible to this enzyme, while intact mRNAs carry 5′-triphosphates and are resistant to the degradation. Since it employs a general feature of processed rRNAs, this method is anticipated to work broadly for bacteria and archaea, therefore predicted to be useful for complex microbial communities. However, unexpected low removal of rRNAs from an archaeal *Halorhabdus* sp. by this kit was observed.[6] The authors suspected that some of the processed 5′ ends of archaeal rRNAs may have exposed a hydroxyl group rather than a monophosphate, as also suggested by the poor ligation of archaeal *Sulfolobus solfataricus* rRNAs to adapters that require an exposed 5′-monophosphate.[15]

The effectiveness and fidelity of this method can be affected by the presence of partially degraded mRNA in crude extracts.[6] In bacteria and archaea, many mRNAs have short half-lives and degrade in minutes after transcription.[16] Continuous transcription is needed to maintain translation. Because of the rapid turnover, many extracted mRNAs may be partially degraded. One of the mRNA degradation mechanisms in bacteria starts with the conversion of 5′-triphosphates to monophosphates, which facilitates subsequent internal cleavage of mRNA by RNase E.[17] Therefore, a fraction of mRNAs is monophosphorylated and fragmented, and thus susceptible to the 5′-monophosphate-specific exonuclease digestion. Indeed, mRNAs with short half-lives and high turnover rates, presumably partially degraded mRNAs, were preferentially lost after applying this kit.[6] Such mRNA degradation not only compromised mRNA fidelity but also counteracted the rRNA removal effect.[6] Conceivably, this enzymatic treatment could provide a useful snapshot of stable full-length mRNA; theoretically, it would be useful to identify transcription start sites because to its enrichment for primary transcripts with 5′-triphosphates.

RNase H DIGESTION

Another enzymatic rRNA elimination method utilizes RNase H that specifically degrades the RNA in DNA:RNA hybrids but not single-stranded RNA.[18] This approach was applied to profile *Staphylococcus aureus* gene expression.[19] Total RNA was first subjected to reverse transcription with a mix of primers specific to 16S and 23S rRNAs. Subsequently, RNase H was added to degrade the RNA fraction of the cDNA:RNA duplex, followed by a DNase I treatment to degrade the cDNA fraction. Therefore, complete RNase H and DNase I digestions are required for the effectiveness of this method. Since the specificity is affected by primer sequences used in reverse transcription, potential limitations and biases associated with primer targeting range are anticipated. In addition, as the reverse transcriptase falls off templates with a certain probability, the longer the transcript is, the less likely its full length is covered. Therefore, the 5′ ends of rRNA, especially 23S rRNA, may not be effectively removed. A number of primers targeting multiple sites across rRNA molecules are needed to improve effectiveness.

SUBTRACTIVE HYBRIDIZATION WITH rRNA-SPECIFIC PROBES

Subtractive hybridization employs antisense sequences of rRNAs as specific probes. After hybridization, rRNA:probe hybrids bind to beads, allowing their subtraction from the solution.[20] Subtractive hybridization is the most commonly used method so far, partly owning to the availabilities of several commercial kits (Table 3.1).

The MICROBExpress kit from Ambion (Austin, TX) is the most widely used commercial kit in depleting rRNA from bacteria. One end of the capture oligonucleotide probes targets conserved regions of 16S and 23S rRNAs, and the other end contains supplementary sequence to a secondary probe coated on magnetic beads. This probe configuration allows the capture of rRNA:probe hybrids to the beads after two sequential hybridizations. A major strength of this kit is the great rRNA removal fidelity. The correlations of mRNA transcript abundance before and after rRNA removal by MICROBExpress were usually higher than other evaluated methods.[6–8] For example, only a small number (2–3%) of transcripts exhibited greater than twofold change in expression relative to the sample without rRNA removal, suggesting that probe cross-hybridization to mRNA was minimal.[6] However, the removal efficiency is highly dependent on organism(s) present in the sample, since the probes are only compatible with a subset of bacteria and the entire archaeal domain is not targeted. In some applications, two rounds of MICROBExpress were applied sequentially on metatranscriptome samples,[21] presumably to improve mRNA enrichment. However, upon evaluation, two rounds did not produce a significant improvement in rRNA depletion over one round, suggesting that rRNAs with target sites were efficiently removed in the first round and additional round is unnecessary.[6] In addition to the limited target range, the success of this kit is also affected by RNA integrity, as demonstrated by the reverse correlation between rRNA removal and RNA integrity.[6] Because of the very small number of probes employed by this kit, degradation and fragmentation increase the proportion of rRNA fragments without the conserved target sites, and therefore negatively affect the removal efficiency.

The more recently introduced Ribo-Zero kit (Epicentre) utilizes rRNA-specific biotinylated probes that can be subsequently captured by streptavidin-coated magnetic beads.[22] Compared with MICROBExpress, Ribo-Zero offers rRNA removal from a broader range of Gram-positive and Gram-negative bacteria. From the probe compatibility table provided by the manufacture, in addition to 16S and 23S, Ribo-Zero also removes 5S rRNA in some cases, and several archaeal species are also targeted. In the comprehensive evaluation conducted by Giannoukos *et al.*,[7] Ribo-Zero clearly outperformed four other kits/methods on bacteria with different GC contents, consistently reducing rRNA to less than 1% of total reads, while maintaining a high fidelity comparable to that offered by MICROBExpress. When applied to real environmental samples from human feces, Ribo-Zero was able to reduce rRNA to less than 5% of total reads.[7] In addition, Ribo-Zero is also effective for RNA of low integrity, as confirmed by using a highly fragmented RNA sample.[7] This improvement is likely because of higher numbers of proprietary probes designed along rRNA molecules. The effectiveness of Ribo-Zero on metatranscriptome samples from diverse environments remains to be examined. Notably, it is suggested that its application on low-input RNA (e.g., below microgram quantities) should be avoided, as the effectiveness and fidelity were greatly compromised.[7] Therefore, this kit is probably not suitable for environmental samples with low RNA yields.

In general, subtractive hybridizations using commercial kits can achieve high transcript fidelity but suffer from the limited target range offered by generic probes. Therefore, sample-specific subtraction strategies have been applied using customized probes tailored for individual organisms or samples. In fact, prior to commercial kits, early versions of subtractive hybridization are organism specific, using biotin-labeled full-length or near full-length rRNA genes cloned into a plasmid[23] or antisense rRNA *in vitro* transcribed from cloned rRNA genes.[24,25] More recently, Li *et al.*[26] developed an organism-specific subtraction system.

They first designed a computer program to select probes specific for 16S and 23S rRNA of the target organism without cross-hybridization to its mRNAs. The probe sequences were then polymerase chain reaction (PCR)-amplified from cDNA, cloned into a plasmid system and *in vitro* transcribed to generate biotin-labeled antisense rRNA probes. They tested this system on *Mycobacterium smegmatis* and obtained significant rRNA reduction with minimal impact on mRNA. Theoretically, this system can be applied to monocultures or simple communities with known reference genome information.

For complex communities, Stewart *et al.*[11] developed a community-specific rRNA subtraction protocol potentially applicable to many different metatranscriptome samples. This method starts with extracting community DNA and RNA in parallel from a same sample. Near full-length 16S and 23S rRNA genes were PCR-amplified from the DNA using bacterial, archaeal, and eukaryotic general primers, respectively. Reverse primers contain T7 RNA polymerase promoter sequence, allowing subsequent *in vitro* transcription to generate biotin-labeled antisense rRNAs to subtract community rRNA. Ideally, probes generated from community RNA are more representative to rRNA composition than from DNA. However, as reverse transcription is less efficient for long transcripts (e.g., 23S rRNA), Stewart *et al.*[11] chose DNA as templates in order to synthesize nearly full-length antisense rRNA probes to maximize probe coverage along rRNA molecules for their effectiveness on fragmented rRNA. When applied on marine metatranscriptome samples, this protocol successfully reduced rRNA to an average of 36% of total reads.[11] As a single primer set may not amplify all phylogenetic groups present in a complex sample, and the secondary structure of near full-length rRNAs may decrease hybridization efficiency, this protocol was later modified by Kukutla *et al.*[27] and tested on a mosquito gut metatranscriptome. Probes generated from the four combinations of two forward and two reverse primers were pooled to maximize phylogenetic coverage. Some primer combinations generate shorter probes, presumably less prone to form secondary structure, and thus may facilitate the probes–rRNA hybridization. Despite the success, the community-specific subtraction method has not been widely applied, mainly because of the time-consuming procedures, requiring more skills and individual optimization, compared with commercial kits.

BIASED 3′ POLYADENYLATION AGAINST rRNA

Amara and Vijaya[28] observed selective polyadenylation of *E. coli* mRNA but not rRNA in intact polysomes using a yeast-derived poly(A) polymerase, suggesting that the 3′ ends of rRNAs are sterically blocked from polyadenylation when rRNAs are in complex with ribosomal proteins. Based on this observation, Wendisch *et al.*[29] developed a bacterial mRNA enrichment method using biased polyadenylation toward mRNA followed by purification with oligo(dT) chromatography. They tested this method on *E. coli* and obtained increased mRNA signals with similarity in transcript profiles when compared with results obtained from total RNA. Besides oligo(dT) chromatography purification, polyadenylated mRNAs can also be selectively synthesized to cDNA using oligo(dT) primers, and *in vitro* transcribed with T7 RNA polymerase to further boost the mRNA enrichment effect; this was applied by Frias-Lopez *et al.*[30] to enrich mRNA from nanogram quantities of marine microbial community RNA. They reduced rRNA to 53% of the total sequences, and maintained relative abundance of most mRNAs, compared with unamplified transcriptome profiles. As the authors suggested, the enrichment was not only because of preferential polyadenylation but also likely arose from the high degree of rRNA secondary structure, which prevented their efficient amplification.[30] However, as Wilhelm and Landry[31] pointed out, bias against 5′ ends and especially against long transcripts was expected when priming with oligo(dT) primer as compared with random oligomers in reverse transcription, since reverse transcriptase is well known to fall off templates as it progresses.

SELECTIVE PRIMING IN cDNA SYNTHESIS

Usually, cDNA synthesis is primed with random oligomers (most frequently hexamers) for "unbiased" transcriptome coverage. Theoretically, using a subset of these entirely random oligomers that discriminates against rRNAs, one can selectively synthesize cDNA from mRNA. Based on the observation that many bacterial and archaeal rRNA sequences are richer in GC compared with mRNA and exhibit higher frequencies of TTTT, Gonzalez and Robb[32] applied HDTTTT and DHTTTT hexamers to prime cDNA synthesis from *Methanocaldococcus jannaschii* and *Pyrococcus horikoshii* and improved their mRNA detection by PCR. This method is easy to conduct in the laboratory and does not require high RNA inputs. Although the primer design was intended to be "universal" for bacteria and archaea, the differential occurrence of TTTT in rRNA versus in non-rRNA broadly varies among microorganisms, causing variation in removal effectiveness and associated bias, especially against GC-rich regions in a transcriptome and GC-rich organisms in a community. Therefore, selective primers with higher specificity were later designed computationally. By aligning all possible hexamers to rRNA and non-rRNA genes, Armour *et al.*[33] selected a subset of hexamers without perfect match to rRNA while providing sufficient coverage to non-rRNA (referred to as "not-so random primers"). This approach was applied to *Rhodopseudomonas palustris*, and 925 out of the total 4096 possible hexamers were selected.[34] However, nonuniform coverage even within a transcript was observed, due to reduced randomness in priming cDNA synthesis.[34] Therefore, this method is not suitable for applications such as transcription start site detection and operon relationship determination, where coverage uniformity is required.

A commercial kit with generic selective primers, Ovation Prokaryotic RNA-Seq System was developed by NuGEN (San Carlos, CA).[35] The not-so random primers were designed against a sequence collection from bacterial and archaeal strains representing major phyla, therefore predicted to be effective for a wide range of prokaryotes and even for partially degraded RNA. However, when tested on different bacterial monocultures, the effectiveness of this kit widely varied.[7,8] Another feature of this kit is that the double-stranded cDNA can be further amplified by PCR, which allows sequencing from very low quantities of input RNA. However, this meantime may introduce biases well known to exponential amplification, therefore undermining quantitativeness. For example, when applied on a *Burkholderia* sp., the correlation of mRNA abundance between transcriptomes with and without rRNA removal, as indicated by R^2, was only 0.51, much lower than 0.96 obtained by MICROBExpress.[8] In another evaluation, a total of five methods, including Ovation were tested.[7] Subtractive hybridization kits achieved the highest fidelity (R^2 = 0.90–0.97 for MICROBExpress, and R^2 = 0.88–0.95 for Ribo-Zero), followed by cDNA library normalization method discussed below (R^2 = 0.13–0.98). Ovation obtained much lower fidelity

(R^2 = 0.43–0.53), only higher than the mRNA-ONLY kit, for which R^2 was 0.04–0.42.

cDNA LIBRARY NORMALIZATION

rRNA can also be eliminated after cDNA synthesis through cDNA library normalization, which had been previously used to "equalize" transcripts in order to recover rare ones.[36] Double-stranded cDNA is first denatured at an elevated temperature and then allowed to re-anneal at a lower temperature. Because the re-annealing rate of a cDNA is proportional to the square of its concentration, cDNAs from abundant transcripts (mostly rRNA and tRNA) return to double-stranded forms faster than from less-abundant transcripts. The re-annealing is terminated after an appropriate time period so that most rRNA-deriving cDNAs are in the double-stranded form, while cDNAs from mRNAs remain as single stranded. The double-stranded cDNA can then be removed from single-stranded ones through enzymatic degradation with a duplex-specific nuclease (DSN)[9] or through physical separation with hydroxyapatite chromatography.[37] For example, applying cDNA library normalization with DSN on *E. coli* reduced rRNA representation in the total sequences from 94% to 26%, while preserving mRNA relative abundance, with only 10 most abundant mRNA transcripts exhibiting a small reduction after the DSN treatment.[9] As the treatment occurs after cDNA library construction, the RNA input can be lowered to submicrogram quantities (e.g., 200–300 ng[9,37]).

However, the robustness of this method needs further investigation. As the re-annealing rate is concentration dependent, a fixed re-annealing time might not work for samples with varied concentrations of rRNA and mRNA, and this might partly explain the observed variation of rRNA removal efficiency among samples.[9] Furthermore, variation is also associated with genome GC contents, as demonstrated by the efficient removal with high fidelity on the low- and medium-GC bacteria but failure on the high-GC bacterium tested.[7] This suggested that the high GC fraction of mRNA may anneal faster or form hairpins, thus was also degraded by DSN, leading to its under representation.[7] Another adverse factor is the fact that within a transcriptome, mRNA abundance differs by several orders of magnitude. Although facilitating the detection of rare transcripts, the equalization on the contrary, reduces the dynamic range of mRNA abundance, and thus compromises quantitativeness. The effectiveness and fidelity may become even more difficult to achieve for environmental samples with communities members differing in GC content and relative abundance. Conceivably, rRNA from low-abundance organisms and highly expressed mRNA from dominant organisms may be at the same concentration range, making it difficult to differentiate rRNA and mRNA.

COMBINATION OF METHODS

Because of the limitation of each method, one method alone does not offer sufficient rRNA removal in some cases, particularly for challenging organisms or complex microbial communities. Therefore, researchers have tried combining different methodologies to improve mRNA enrichment. For example, 5′-monophosphate-dependent exonuclease digestion followed by subtractive hybridization was applied to coastal water metatranscriptome samples, and it decreased rRNA sequences to 37% of total reads.[38] When systematically evaluated, the combinations of these two methods provided more efficient rRNA removal than used alone, but produced much greater bias in mRNA abundance regardless of the order in which they were applied, and therefore should be avoided.[6]

In contrast to the increased bias by combining subtractive hybridization and exonuclease digestion, Peano *et al.*[8] observed an interesting phenomenon when combining hybridization and selective priming. When removing rRNA from a high GC *Burkholderia* sp., MICROBExpress alone, although offering a very high fidelity, was not efficient, while Ovation alone reduced rRNA to 61% of total reads, but introduced a significant bias in mRNA. Interestingly, applying MICROBExpress followed by Ovation not only reduced rRNA to 54% of total reads but also significantly increased fidelity as compared to using Ovation alone, suggesting that the application of MICROBExpress prior to Ovation made the latter less prone to bias.[8] The mechanism for such bias reduction is unclear and needs further investigation for its wide application on high GC microorganisms.

In general, combination of methods might be able to improve rRNA removal efficiency in some cases, but the additional treatment and purification procedures cause more material loss, and the extended sample processing time increases the risk of mRNA degradation, since mRNA is less stable than rRNA and tRNA. These factors need to be considered in method combination.

CONCLUSIONS AND PERSPECTIVES

Progress has been made to remove rRNA from total RNA in microbial transcriptomic analysis. Each removal method has its own strengths and limitations. The choice of method depends not only on samples (e.g., GC content, genome complexity, phylogenic composition, and RNA quantity and quality) but also on applications, which may differ in the requirements for quantitativeness in mRNA abundance, uniformity in transcript coverage, representation of rare transcripts, etc. Among methods compared so far, subtractive hybridization is the most widely used and generates the least bias. However, customized probes need to be synthesized if samples are not targeted by commercial probes, and its application is limited to samples with microgram quantities. Methods such as biased polyadenylation, selective priming, and cDNA library normalization are less limited by sample phylogenetic composition and are suitable for low-quantity RNA, but need to be used with caution for potential biases. For low-yield environmental samples where the rRNA content is expected not as high as in fast-growing laboratory cultures, sequencing without rRNA removal might be a viable choice to avoid degradation of mRNA because of extensive sample processing. With further sequencing throughput increase and cost drop, rRNA depletion may become less cost-effective and less beneficial, when considering the labor and potential biases introduced because of additional handling. For example, it was suggested that with current ultra-high throughputs such as from HiSeq 2000, rRNA depletion may not be necessary for analyzing single-species monocultures.[39] Conceivably, rRNA removal will become less crucial in performing RNA-Seq analysis of complex community samples in the future.

REFERENCES

1. Moran MA. Metatranscriptomics: eavesdropping on complex microbial communities. Microbe 2009;4(7):329–35.

2. Sorek R, Cossart P. Prokaryotic transcriptomics: a new view on regulation, physiology and pathogenicity. Nat Rev Genet 2010;11:9–16.

3. Filiatrault MJ. Progress in prokaryotic transcriptomics. Curr Opin Microbiol 2011;14(5):579–86.
4. Neidhardt FC, Umbarger HE. Chemical Composition of *Escherichia coli*. 2nd ed. Washington, D.C.: ASM Press; 1996.
5. Karpinets TV, Greenwood DJ, Sams CE, Ammons JT. RNA: protein ratio of the unicellular organism as a characteristic of phosphorous and nitrogen stoichiometry and of the cellular requirement of ribosomes for protein synthesis. BMC Biol 2006;4(1):30.
6. He S, Wurtzel O, Singh K, Froula JL, Yilmaz S, Tringe SG, et al. Validation of two ribosomal RNA removal methods for microbial metatranscriptomics. Nat Methods 2010;7(10):807–12.
7. Giannoukos G, Ciulla D, Huang K, Haas BJ, Izard J, Levin JZ, et al. Efficient and robust RNA-seq process for cultured bacteria and complex community transcriptomes. Genome Biol 2012;13:R23.
8. Peano C, Pietrelli A, Consolandi C, Rossi E, Petiti L, Tagliabue L, et al. An efficient rRNA removal method for RNA sequencing in GC-rich bacteria. Microbial Inform Experiment 2013;3(1):1.
9. Yi H, Cho Y-J, Won S, Lee JE, Yu HJ, Kim S, et al. Duplex-specific nuclease efficiently removes rRNA for prokaryotic RNA-seq. Nucleic Acids Res 2011;39(20):e140.
10. Zhao J, Hyman L, Moore C. Formation of mRNA 3′ ends in eukaryotes: mechanism, regulation, and interrelationships with other steps in mRNA synthesis. Microbiol Mol Biol Rev 1999;63(2):405–45.
11. Stewart FJ, Ottesen EA, DeLong EF. Development and quantitative analyses of a universal rRNA-subtraction protocol for microbial metatranscriptomics. ISME J 2010;4(7):896–907.
12. McGrath KC, Thomas-Hall SR, Cheng CT, Leo L, Alexa A, Schmidt S, et al. Isolation and analysis of mRNA from environmental microbial communities. J Microbiol Methods 2008;75(2):172–6.
13. Azarani A, Hecker KH. RNA chromatography under thermally denaturing conditions: analysis and quality determination of RNA. Appl Note 2000;116:1–4.
14. Castro TLP, Seyffert N, Ramos RTJ, Barbosa S, Carvalho R, Pinto AC, et al. Ion Torrent-based transcriptional assessment of a Corynebacterium pseudotuberculosis equi strain reveals denaturing high-performance liquid chromatography a promising rRNA depletion method. Microbial Biotechnol 2013;6(2):168–77.
15. Wurtzel O, Sapra R, Chen F, Zhu Y, Simmons BA, Sorek R. A single-base resolution map of an archaeal transcriptome. Genome Res 2010;20(1):133–41.
16. Rauhut R, Klug G. mRNA degradation in bacteria. FEMS Microbiol Rev 1999;23(3):353–70.
17. Jiang X, Belasco JG. Catalytic activation of multimeric RNase E and RNase G by 5′-monophosphorylated RNA. Proc Natl Acad Sci USA 2004;101(25):9211–6.
18. Hausen P, Stein H, Ribonuclease H. An enzyme degrading the RNA moiety of DNA-RNA hybrids. Eur J Biochem/FEBS 1970;14(2):278–83.
19. Dunman PM, Murphy E, Haney S, Palacios D, Tucker-Kellogg G, Wu S, et al. Transcription profiling-based identification of Staphylococcus aureus genes regulated by the agr and/or sarA loci. J Bacteriol 2001;183(24):7341–53.
20. Pang X, Zhou DS, Song YJ, Pei D, Wang J, Guo Z, et al. Bacterial mRNA purification by magnetic capture-hybridization method. Microbiol Immunol 2004;48(2):91–6.
21. Shrestha PM, Kube M, Reinhardt R, Liesack W. Transcriptional activity of paddy soil bacterial communities. Environ Microbiol 2009;11(4):960–70.
22. Sooknanan R, Pease J, Doyle K. Novel methods for rRNA removal and directional, ligation-free RNA-seq library preparation. Nat Methods Appl Notes 2010;7(10).

23. Robinson KA, Robb FT, Schreier HJ. Isolation of maltose-regulated genes from the hyperthermophilic archaeum, Pyrococcus furiosus, by subtractive hybridization. Gene 1994; 148(1):137–41.

24. Su CL, Sordillo LM. A simple method to enrich mRNA from total prokaryotic RNA. Mol Biotechnol 1998;10(1):83–5.

25. Plum G, Clark-Curtiss JE. Induction of Mycobacterium avium gene expression following phagocytosis by human macrophages. Infect Immun 1994;62(2):476–83.

26. Li S-K, Zhou J-W, Yim AK-Y, Leung AKY, Tsui SKW, Chan TF, et al. Organism-Specific rRNA capture system for application in next-generation sequencing. PLoS One 2013;8(9):e74286.

27. Kukutla P, Steritz M, Xu J. Depletion of ribosomal RNA for mosquito gut metagenomic RNA-seq. JoVE *(Journal of Visualized Experiments)* 2013;(74):e50093.

28. Amara RR, Vijaya S. Specific polyadenylation and purification of total messenger RNA from Escherichia coli. Nucleic Acids Res 1997;25(17):3465–70.

29. Wendisch VF, Zimmer DP, Khodursky A, Peter B, Cozzarelli N, Kustu S. Isolation of Escherichia coli mRNA and comparison of expression using mRNA and total RNA on DNA microarrays. Anal Biochem 2001;290(2):205–13.

30. Frias-Lopez J, Shi Y, Tyson GW, Coleman ML, Schuster SC, Chisholm SW, et al. Microbial community gene expression in ocean surface waters. Proc Natl Acad Sci USA 2008;105(10):3805–10.

31. Wilhelm BT, Landry JR. RNA-Seq-quantitative measurement of expression through massively parallel RNA-sequencing. Methods 2009;48(3):249–57.

32. Gonzalez JM, Robb FT. Counterselection of prokaryotic ribosomal RNA during reverse transcription using non-random hexameric oligonucleotides. J Microbiol Methods 2007;71(3):288–91.

33. Armour CD, Castle JC, Chen R, Babak T, Loerch P, Jackson S, et al. Digital transcriptome profiling using selective hexamer priming for cDNA synthesis. Nat Methods 2009;6(9):647–9.

34. Hirakawa H, Oda Y, Phattarasukol S, Armour CD, Castle JC, Raymond CK, et al. Activity of the Rhodopseudomonas palustris p-coumaroyl-homoserine lactone-responsive transcription factor RpaR. J Bacteriol 2011;193(10):2598–607.

35. Head SR, Komori HK, Hart GT, Shimashita J, Schaffer L, Salomon DR, et al. Method for improved Illumina sequencing library preparation using NuGEN Ovation RNA-Seq system. Biotechniques 2011;50(3):177–80.

36. Ko MSH. An "equalized cDNA library" by the reassociation of short double-stranded cDNAs. Nucleic Acids Res 1990;18(19):5705–11.

37. VanderNoot VA, Langevin SA, Solberg OD, Lane PD, Curtis DJ, Bent ZW, et al. cDNA normalization by hydroxyapatite chromatography to enrich transcriptome diversity in RNA-seq applications. Biotechniques 2012;53(6):373.

38. Poretsky RS, Hewson I, Sun S, Allen AE, Zehr JP, Moran MA. Comparative day/night metatranscriptomic analysis of microbial communities in the North Pacific subtropical gyre. Environ Microbiol 2009;11(6):1358–75.

39. Haas B, Chin M, Nusbaum C, Birren B, Livny J. How deep is deep enough for RNA-Seq profiling of bacterial transcriptomes? BMC Genomics 2012;13(1):734.

CHAPTER 4

High-Throughput Sequencing as a Tool for Exploring the Human Microbiome

Mathieu Almeida and Mihai Pop

INTRODUCTION

The advent of massively parallel sequencing technologies has dramatically broadened the scope of sequencing applications to new biological domains. In metagenomics, in particular, the combined DNA of entire microbial communities can now be sequenced effectively and cheaply, leading to the development of new analytical methodologies that leverage the new types of data being generated. It is important to note that analyses of whole-community DNA had already been performed long before the advent of new sequencing technologies, as evidenced by Ed de Long's pioneering work in this field that identified bacterial rhodopsin in marine bacteria,[1] and the first high-throughput explorations of acid mine drainage,[2] ocean,[3] and human distal gut[4] microbial communities, all performed with the Sanger technology. The new sequencing technologies have simply enabled a much broader and cost-effective use of sequencing in these studies.

Here we will discuss the main issues related to the computational analysis of the resulting data. We will specifically focus on approaches that analyze the entire community DNA (commonly termed *metagenomic experiments*) in contrast to approaches that focus on specific genes within the community (usually, molecular community surveys based on targeted sequencing of the 16S rRNA gene).

Before we proceed, we would like to note that the promise of metagenomic studies is the ability to provide a glimpse at the majority of organisms that cannot be readily grown in a laboratory. Even in heavily studied environments, such as the human intestinal tract, it is estimated that less than 30% of all microbes have been isolated.[5] High-throughput sequencing provides a way to access the genomic content of the uncultured

Metagenomics for Microbiology. http://dx.doi.org/10.1016/B978-0-12-410472-3.00004-X

members of the community. Note, however, that this view is distorted by the inability of current analytical tools to fully reconstruct genomes from the relatively short sequences currently generated (ranging from ~100 bp for Illumina to ~500 bp for Roche/454 or IonTorrent), and even to untangle the multiple closely related genomes possibly found in an environmental mixture. Simply put, a metagenomic shotgun sequencing experiment loses a large part of the information contained in the genomic mixture, information that cannot be fully recovered through computational means. In addition, the sequencing data alone provide just hints into the biological functions of individual organisms or whole communities. In other words, the approaches we describe below and their results should be considered as just a first (and very important) step in our attempts to understand the structure and function of microbial communities. Rather than providing conclusive answers, these methods generate new hypotheses that will form the basis of future experimental studies.

Furthermore, an important emerging area of research is the incorporation of orthogonal types of data, such as gene or protein expression data, which are starting to be generated for metagenomic communities. Such methods are not discussed in our review as they are still in their infancy.

ORGANIZING METAGENOMIC READS BY MAPPING ONTO REFERENCE GENOMES

Although we expect that many of the organisms found in an environmental mixture have never been isolated and are thus absent from public databases, valuable insights can be obtained by comparing metagenomic sequences to currently available datasets. The reference provided by already sequenced genomes arguably provides the most reliable substrate for identifying the taxonomic origin of individual metagenomic reads and also enables the exploration of the genome structure of organisms closely related to previously sequenced genomes. The latter information is difficult to obtain from reads alone as will be detailed further. The value of reference genomes as a substrate for the analysis of metagenomic data is well recognized in the community, and has led to the initiation of efforts aimed at sequencing the genomes of isolate organisms from culture collections,[6] as well as for extracting and sequencing organisms of interest from environmental mixtures.[7] It is important to note that the number of organisms already isolated but not yet sequenced is substantial, and sequencing these organisms is a valuable first step in constructing a relevant reference collection for metagenomic studies. Such efforts are, however, stymied by the complex and time-intensive procedures necessary for submitting a genome sequence to public repositories – as a result, the increase in the number of publicly available bacterial genomes is far slower than would be expected given the dramatic improvements in sequencing technologies.

The use of reference genomes for the analysis of metagenomic data is further complicated by the sheer size of the data being analyzed. Thousands of complete or nearly complete genomes are currently publicly available and we can expect this number to grow to tens or even hundreds of thousands of genomes in the not too distant future. Current alignment algorithms, such as Bowtie,[8] BWA,[9] SOAP,[10] and others, cannot efficiently search such large collections. Furthermore, recent work on the development of new alignment tools specifically targeted at genome collections is primarily focused on human populations and cannot be easily extended to microbial collections. There is a critical need in the community for novel approaches that can search the massive reference databases that will soon be available.

Despite these limitations, reference genomes have been effectively used in the analysis of metagenomic data. In a recent analysis of 39 human fecal samples,[11] metagenomic sequences (comprising data from a mixture of technologies) were mapped to a collection of 1506 reference genomes in order to quantify the abundance of these genomes (or close relatives thereof) within the metagenomic samples. The resulting abundance matrices were used to compare and cluster the individual samples, analysis that revealed a broad stratification into three major clusters, termed enterotypes. This analysis also revealed that only approximately 40% of the reads could be mapped to reference genomes, as expected given the estimated fraction of gut microbes that have not yet been isolated.

In another recent example, sequences from 252 stool samples were mapped onto a collection of 1497 genomes in order to identify individual-specific polymorphisms (focused on single-nucleotide polymorphisms) distinguishing the metagenomic organisms from their previously sequenced relatives.[12] This study revealed that individual-specific variation patterns were stable over long periods of time in healthy individuals, suggesting that the microbial genotype of a person may represent an identifiable marker similar to the host genotype. Similar to the above-mentioned study, once again only 40% of the reads on average could be aligned to the reference genomes.

The alignment of metagenomic sequences to reference genomes can also provide insights into the specific adaptations of organisms in specific samples. For example, mapping sequences from a healthy oral microbiome against the sequenced genome of *Actinomyces naeslundii*, genome originally isolated from a diseased patient, revealed genomic differences possibly associated with pathogenicity, such as genes related to mercury resistance and drug transporters found in the potentially pathogenic *Actinomyces* strain but absent from the metagenomic samples.[13]

TAXONOMIC CLASSIFICATION/BINNING

As described above, methods that rely on mapping reads to previously sequenced genomes fail to characterize large fractions of microbial populations as many organisms have yet to be isolated and sequenced. Existing alignment algorithms can only discover very close relationships and can only be used to analyze the environmental organisms most closely related to genomes in public databases. More distant relationships can be inferred through the use of machine-learning techniques in a process called taxonomic binning or taxonomic classification. These tools attempt to assign each read to a taxonomic "bin" generally approximating a broad taxonomic group such as a genus or a family. One of the earliest tools in this field, MEGAN,[14] relied on simple blast searches (more sensitive than current short-read aligners) to identify matches against a database of sequences with known taxonomic origin. Other approaches focus on the k-mer (short DNA patterns, usually three to four base pairs in length) profiles of sequences that are matched, using machine-learning techniques, against precomputed databases constructed from known genomes. DNA composition is broadly consistent across taxonomic groups[15] thus providing a useful classification signature even when sequence similarity cannot be detected through alignment. Among such methods are tools based on self-organizing maps,[16] interpolated Markov models – Phymmbl,[17] naïve Bayes classifiers – NBC,[18] and support vector machines – Phylopythia.[19]

Compositional-based approaches such as those described above can be stymied by genomic regions with unusual DNA compositions or lateral gene transfer. These limitations can be addressed by focusing on just specific genes that are considered phylogenetically informative, that is, their composition correlates with the evolutionary history of the organisms. Tools that leverage this information include Amphora,[20] Metaphyler,[21] MetaPhlAn,[22] and mOTU.[23]

STRUCTURING METAGENOMIC SHORT READS INTO A GENE CATALOG BY *DE NOVO* ASSEMBLY

Knowledge of the broad taxonomic origin of metagenomic sequences is not sufficient for understanding the function of microbes within a community, as closely related genomes may differ in clinically relevant functions (e.g., differences between commensal and pathogenic *Escherichia coli* strains). Furthermore, human gut communities were shown to exhibit relatively stable functional profiles despite highly divergent taxonomic compositions.[24,25] Reconstructing genes and even genomes from metagenomic mixtures is an important first step toward a better characterization of their functional profile, although some analyses are possible even when starting with the reads alone.[26]

Genome assembly is a difficult task even for isolate genomes, and its difficulty is compounded in metagenomic samples for three main reasons: (i) low-abundance organisms cannot be effectively assembled because of lack of coverage; (ii) wide differences in abundance/coverage between community members makes it difficult to identify genomic repeats; and (iii) true differences between closely related organisms cannot be easily distinguished from sequencing errors. Despite these challenges, early metagenomic studies relied on assembly tools developed for isolate genomes such as Celera Assembler[27] in the first gut microbiome[4] or SOAPdeNovo[28] in the MetaHit[5] and Human Microbiome Project (HMP)[29] studies.

Recently, a number of tools have been developed that target the specific characteristics of metagenomic data, including Meta-IDBA,[30]

Meta-Velvet,[31] and integrated pipelines that include assembly as well as downstream analyses such as gene finding and taxonomic classification – MOCAT[32] and MetAMOS.[33]

It is important to note that the ability to effectively sequence and reconstruct a substantial part of a complex microbial mixture is by no means obvious. Generating sufficient sequencing depth to ensure the data can be assembled might be prohibitively expensive, and even if sufficient data are generated, the data may not be easily analyzed computationally. The analysis of the data generate in the HMP revealed that, at least for fecal samples, these concerns are not warranted. Specifically, two lanes of paired-end reads from an Illumina GA2 instrument (currently equivalent to a single lane of an Illumina HiSeq experiment) suffice to reconstruct a substantial fraction of the gut microbiome, and the addition of sequencing depth does not significantly improve the quality of the reconstruction.[25] In other host-associated communities, however, the high level of human DNA contamination dramatically reduces the effective sequencing depth available, highlighting the need for the development of strategies for enriching the microbial component of the samples.[25] In addition, highly complex communities, such as those found in soil, continue to present a significant analytical challenge.[34]

Although the goal of assembly is to reconstruct entire genomes, the output of metagenomic assemblers is highly fragmented and requires additional analyses to identify the sets of contigs that belong to a same genome. Compositional-based methods (such as those used for taxonomic classification) and depth of coverage information can be used for this purpose,[2] although more elaborate analyses of the data may require manual inspection of the assembly results, as performed, for example, in an analysis of the strain structure in a *Citrobacter* population within the developing gut microbial communities in infants.[35]

Because of the lack of contiguity and general complexity of assembled metagenomic data, most studies so far have focused on characterizing the genic content of the data rather than on the reconstruction of individual organisms. Such analyses have revealed the tremendous diversity of genes harbored in microbial ecosystems: nonredundant gene catalogs comprised 6.1 million genes in an oceanic microbiome,[36] 3.3 million genes in the human gut data generated by the MetaHit project,[5] and

5 million genes in the data generated by the HMP.[29] These gene catalogs have dramatically expanded the known universe of genes – for example, the Global Ocean Survey data alone significantly exceeded the cumulative size of the National Center for Biotechnology Information (NCBI) nonredundant database when this dataset was created.[36] In the context of gut microbial communities, it appears that the above-mentioned gene catalogs capture a significant fraction of the collective functional content of gut microbiota across the human population: 90% of the genes identified in the MetaHit projects are also found in the HMP catalog, as well as within a similar catalog constructed from Chinese subjects.[37]

CLUSTERING METAGENOMIC GENE CATALOGS

The sheer size of metagenomic gene catalogs (ranging in the millions of genes, as described above) makes it difficult to analyze and interpret the resulting data. A recently proposed solution involves the use of gene abundances across multiple samples (as estimated, e.g., by mapping metagenomic reads to the gene catalog) to identify genes with correlated abundance patterns.[37,38] Genes with highly correlated abundance patterns can be inferred to originate from a same chromosome, thus allowing one to reconstruct virtual genome clusters. Note, however, that high correlation can also be expected because of symbiotic or mutualistic interactions between environmental members, factor that can lead to the false aggregation of genes from distinct organisms. On the contrary, variable genes, such as virulence cassettes or prophage regions, will not cluster with the organism containing them, making it difficult to use the gene clusters to study the specific adaptation of microbes to their environment.

Despite these limitations, such clusters of genes are increasingly used to organize metagenomic data and simplify their analysis and interpretation as evidenced by recent studies associating microbiome composition with type II diabetes[37] and obesity,[39] or tracking the response of the host microbiome during dietary intervention.[40]

Gene clusters can also be used as a framework around which one can reconstruct metagenomic organisms, by iterative recruitment of metagenomic reads mapping to the genes within a cluster followed by the assembly of the resulting sequences.[38] This approach, encoded in the Profile Augmented Metagenomic Assembly (PAMA) pipeline,[38] has allowed the reconstruction of draft genomes from 700 gene clusters. An important question arises in this context, specifically the quality of the reconstructed genomes. This issue is highly relevant to isolate genome as well and has been addressed extensively recently through assembly "bake-offs"[41,42] and the development of validation tools.[43–46] In the context of microbiome studies, the HMP developed a collection of six quality criteria that consider the length and coverage of contigs and scaffolds, as well as the presence of essential functions within the reconstructed genomes.[6] Other proposed measures of quality include the redundancy of genes/functions known to be unique in an organism.[47] Among the 700 genomes described above, 238 genomes passed such a strict set of quality criteria and were deposited into NCBI databases (they can be retrieved by the query "160767[top bioproject]").

ADVANCED METAGENOMIC ANALYSES

Most metagenomic studies described above primarily focused on extensions of analytical methods developed for isolate genomes. The specific characteristics of metagenomic data, as well as the application of metagenomic methods to large collections of samples, make it possible to explore biological questions that cannot be studied in single cultured organisms. Among these are attempts to uncover the network of interactions between community members,[48,49] the exploration of lateral gene transfer events,[50] and the study of the dynamic behavior of microbial ecosystems.[51,52]

A recent exciting development is also the utilization of microbial abundance across multiple samples as a proxy for phenotype, specifically in the context of clustering 16S rRNA gene sequences into operational taxonomic units,[53] allowing for a more precise delineation of taxonomic groups than possible on the basis of sequence data alone.

CONCLUSION

We have outlined above recent computational advances in the analysis of metagenomic data and demonstrated the use of such techniques in beginning to understand the role microbial organisms play in our health and our world. We hope that these results not only highlight the tremendous promise metagenomic studies hold but also show the considerable challenges that need to be overcome in order to fully fulfill this promise. These challenges are in part analytical – new computational approaches and tools need to be developed to fully leverage the information found in metagenomic data; but also logistical in nature, existing infrastructures are not able to cope with the rapid increase in biological knowledge, both in terms of newly sequenced organisms and newly discovered genes. Overcoming these logistical challenges will require substantial cooperation between researchers, funding agencies, and governments.

We would also like to point out that the promise of metagenomics to uncover and characterize novel, previously uncultured, organisms has yet to be fully realized. Initial forays in this direction have already uncovered new organisms from previously unknown phyla[54]; however, they have also revealed the general difficulty of identifying and reconstructing such organisms. Adequate computational tools are simply not available to scientists interested in mining the data for new microbes.

Most of our discussion has focused on bacterial metagenomics; however, all the main concepts apply equally to viral populations. The challenges that have become apparent in the analysis of bacterial metagenomic data are only compounded in viral populations, particularly due to the absence of well-curated and comprehensive reference databases.

Finally, we would like to point out initial forays in understanding the metabolic functions within entire communities[55] that, together with ecological modeling,[56] may lead to the development of a computational modeling infrastructure for metagenomic studies, mirroring successes in this field in isolate genomes.[57]

REFERENCES

1. Beja O, Aravind L, Koonin EV, Suzuki MT, Hadd A, Nguyen LP, et al. Bacterial rhodopsin: evidence for a new type of phototrophy in the sea. Science 2000;289(5486):1902–6.

2. Tyson GW, Chapman J, Hugenholtz P, Allen EE, Ram RJ, Richardson PM, et al. Community structure and metabolism through reconstruction of microbial genomes from the environment. Nature 2004;428(6978):37–43.

3. Rusch DB, Halpern AL, Sutton G, Heidelberg KB, Williamson S, Yooseph S, et al. The Sorcerer II Global Ocean Sampling Expedition: Northwest Atlantic through Eastern Tropical Pacific. PLoS Biol 2007;5(3):e77.

4. Gill SR, Pop M, Deboy RT, Eckburg PB, Turnbaugh PJ, Samuel BS, et al. Metagenomic analysis of the human distal gut microbiome. Science 2006;312(5778):1355–9.

5. Qin J, Li R, Raes J, Arumugam M, Burgdorf KS, Manichanh C, et al. A human gut microbial gene catalogue established by metagenomic sequencing. Nature 2010;464(7285):59–65.

6. Nelson KE, Weinstock GM, Highlander SK, Worley KC, Creasy HH, Wortman JR, et al. A catalog of reference genomes from the human microbiome. Science 2010;328(5981):994–9.

7. Fodor AA, DeSantis TZ, Wylie KM, Badger JH, Ye Y, Hepburn T, et al. The "most wanted" taxa from the human microbiome for whole genome sequencing. PLoS One 2012;7(7):e41294.

8. Langmead B, Trapnell C, Pop M, Salzberg SL. Ultrafast and memory-efficient alignment of short DNA sequences to the human genome. Genome Biol 2009;10(3):R25.

9. Li H, Durbin R. Fast and accurate long-read alignment with Burrows-Wheeler transform. Bioinformatics 2010;26(5):589–95.

10. Li R, Li Y, Kristiansen K, Wang J. SOAP: short oligonucleotide alignment program. Bioinformatics 2008;24(5):713–4.

11. Arumugam M, Raes J, Pelletier E, Le Paslier D, Yamada T, Mende DR, et al. Enterotypes of the human gut microbiome. Nature 2011;473(7346):174–80.

12. Schloissnig S, Arumugam M, Sunagawa S, Mitreva M, Tap J, Zhu A, et al. Genomic variation landscape of the human gut microbiome. Nature 2013;493(7430):45–50.

13. Liu B, Faller LL, Klitgord N, Mazumdar V, Ghodsi M, Sommer DD, et al. Deep sequencing of the oral microbiome reveals signatures of periodontal disease. PLoS One 2012;7(6):e37919.

14. Huson DH, Auch AF, Qi J, Schuster SC. MEGAN analysis of metagenomic data. Genome Res 2007;17(3):377–86.

15. Karlin S, Mrázek J, Campbell AM. Compositional biases of bacterial genomes and evolutionary implications. J Bacteriol 1997;179(12):3899–913.

16. Abe T, Sugawara H, Kanaya S, Ikemura T. A novel bioinformatics tool for phylogenetic classification of genomic sequence fragments derived from mixed genomes of uncultured environmental microbes. Polar Biosci 2006;20:103–12.

17. Brady A, Salzberg SL. Phymm and PhymmBL: metagenomic phylogenetic classification with interpolated Markov models. Nat Methods 2009;6(9):673–6.

18. Rosen GL, Reichenberger ER, Rosenfeld AM. NBC: the Naive Bayes Classification tool webserver for taxonomic classification of metagenomic reads. Bioinformatics 2011;27(1):127–9.

19. McHardy AC, Martin HG, Tsirigos A, Hugenholtz P, Rigoutsos I. Accurate phylogenetic classification of variable-length DNA fragments. Nat Methods 2007;4(1):63–72.

20. Wu M, Eisen JA. A simple, fast, and accurate method of phylogenomic inference. Genome Biol 2008;9(10):R151.

21. Liu B, Gibbons T, Ghodsi M, Treangen T, Pop M. Accurate and fast estimation of taxonomic profiles from metagenomic shotgun sequences. BMC Genomics 2011;12(Suppl 2):S4.

22. Segata N, Waldron L, Ballarini A, Narasimhan V, Jousson O, Huttenhower C. Metagenomic microbial community profiling using unique clade-specific marker genes. Nat Methods 2012;9(8):811–4.

23. Sunagawa S, Mende DR, Zeller G, Izquierdo-Carrasco F, Berger SA, Kultima JR, et al. Metagenomic species profiling using universal phylogenetic marker genes. Nat Methods 2013;10(12):1196–9.

24. Turnbaugh PJ, Hamady M, Yatsunenko T, Cantarel BL, Duncan A, Ley RE, et al. A core gut microbiome in obese and lean twins. Nature 2009;457(7228):480–4.

25. Human Microbiome Project C. Structure, function and diversity of the healthy human microbiome. Nature 2012;486(7402):207–14.

26. Abubucker S, Segata N, Goll J, Schubert AM, Izard J, Cantarel BL, et al. Metabolic reconstruction for metagenomic data and its application to the human microbiome. PLoS Comput Biol 2012;8(6):e1002358.

27. Myers EW, Sutton GG, Delcher AL, Dew IM, Fasulo DP, Flanigan MJ, et al. A whole-genome assembly of Drosophila. Science 2000;287(5461):2196–204.

28. Li Y, Hu Y, Bolund L, Wang J. State of the art de novo assembly of human genomes from massively parallel sequencing data. Hum Genomics 2010;4(4):271–7.

29. Human Microbiome Project C. A framework for human microbiome research. Nature 2012;486(7402):215–21.

30. Peng Y, Leung HC, Yiu SM, Chin FY. Meta-IDBA: a de Novo assembler for metagenomic data. Bioinformatics 2011;27(13):i94–i101.

31. Namiki T, Hachiya T, Tanaka H, Sakakibara Y. MetaVelvet: an extension of Velvet assembler to de novo metagenome assembly from short sequence reads. Proc 2nd ACM Conf Bioinform, Computat Biol Biomed; Chicago, IL: ACM; 2147818, 2011. p. 116–124.

32. Kultima JR, Sunagawa S, Li J, Chen W, Chen H, Mende DR, et al. MOCAT: a metagenomics assembly and gene prediction toolkit. PLoS One 2012;7(10):e47656.

33. Treangen TJ, Koren S, Sommer DD, Liu B, Astrovskaya I, Ondov B, et al. MetAMOS: a modular and open source metagenomic assembly and analysis pipeline. Genome Biol 2013;14(1):R2.

34. Pell J, Hintze A, Canino-Koning R, Howe A, Tiedje JM, Brown CT. Scaling metagenome sequence assembly with probabilistic de Bruijn graphs. Proc Natl Acad Sci USA 2012;109(33):13272–7.

35. Morowitz MJ, Denef VJ, Costello EK, Thomas BC, Poroyko V, Relman DA, et al. Strain-resolved community genomic analysis of gut microbial colonization in a premature infant. Proc Natl Acad Sci USA 2011;108(3):1128–33.

36. Yooseph S, Sutton G, Rusch DB, Halpern AL, Williamson SJ, Remington K, et al. The Sorcerer II Global Ocean Sampling expedition: Expanding the universe of protein families. PLoS Biol 2007;5(3):e16.

37. Qin J, Li Y, Cai Z, Li S, Zhu J, Zhang F, et al. A metagenome-wide association study of gut microbiota in type 2 diabetes. Nature 2012;490(7418):55–60.

38. Nielsen HB, Almeida M, Juncker AS, Rasmussen S, Li J, Sunagawa S, Plichta DR, Gautier L, Pedersen AG, Le Chatelier E, et al. Identification and assembly of genomes and genetic elements in complex metagenomic samples without using reference genomes. *Nat Biotechnol* 2014;32:822–8.

39. Le Chatelier E, Nielsen T, Qin J, Prifti E, Hildebrand F, Falony G, et al. Richness of human gut microbiome correlates with metabolic markers. Nature 2013;500(7464):541–6.

40. Cotillard A, Kennedy SP, Kong LC, Prifti E, Pons N, Le Chatelier E, et al. Dietary intervention impact on gut microbial gene richness. Nature 2013;500(7464):585–8.

41. Earl D, Bradnam K, St John J, Darling A, Lin D, Fass J, et al. Assemblathon 1: a competitive assessment of de novo short read assembly methods. Genome Res 2011;21(12):2224–41.

42. Salzberg SL, Phillippy AM, Zimin A, Puiu D, Magoc T, Koren S, et al. GAGE: A critical evaluation of genome assemblies and assembly algorithms. Genome Res 2012;22(3):557–67.

43. Rahman A, Pachter L. CGAL: computing genome assembly likelihoods. Genome Biol 2013;14(1):R8.

44. Clark SC, Egan R, Frazier PI, Wang Z. ALE: a generic assembly likelihood evaluation framework for assessing the accuracy of genome and metagenome assemblies. Bioinformatics 2013;29(4):435–43.

45. Ghodsi M, Hill CM, Astrovskaya I, Lin H, Sommer DD, Koren S, et al. De novo likelihood-based measures for comparing genome assemblies. BMC Res Notes 2013;6(1):334.

46. Hunt M, Kikuchi T, Sanders M, Newbold C, Berriman M, Otto TD. REAPR: a universal tool for genome assembly evaluation. Genome Biol 2013;14(5):R47.

47. Mende DR, Waller AS, Sunagawa S, Jarvelin AI, Chan MM, Arumugam M, et al. Assessment of metagenomic assembly using simulated next generation sequencing data. PLoS One 2012;7(2):e31386.

48. Friedman J, Alm EJ. Inferring correlation networks from genomic survey data. PLoS Comput Biol 2012;8(9):e1002687.

49. Faust K, Sathirapongsasuti JF, Izard J, Segata N, Gevers D, Raes J, et al. Microbial co-occurrence relationships in the human microbiome. PLoS Comput Biol 2012;8(7):e1002606.

50. Smillie CS, Smith MB, Friedman J, Cordero OX, David LA, Alm EJ. Ecology drives a global network of gene exchange connecting the human microbiome. Nature 2011;480(7376):241–4.

51. Trosvik P, Stenseth NC, Rudi K. Convergent temporal dynamics of the human infant gut microbiota. ISME J 2010;4(2):151–8.

52. Gajer P, Brotman RM, Bai G, Sakamoto J, Schutte UM, Zhong X, et al. Temporal dynamics of the human vaginal microbiota. Sci Transl Med 2012;4(132):132–52.

53. Preheim SP, Perrotta AR, Martin-Platero AM. Distribution-based clustering: using ecology to refine the operational taxonomic unit. Appl Environ Microbiol 2013;79(21):6593–603.

54. Di Rienzi SC, Sharon I, Wrighton KC, Koren O, Hug LA, Thomas BC, et al. The human gut and groundwater harbor non-photosynthetic bacteria belonging to a new candidate phylum sibling to Cyanobacteria. eLife 2013;2:e01102.

55. Greenblum S, Turnbaugh PJ, Borenstein E. Metagenomic systems biology of the human gut microbiome reveals topological shifts associated with obesity and inflammatory bowel disease. Proc Natl Acad Sci USA 2012;109(2):594–9.

56. Trosvik P, Rudi K, Naes T, Kohler A, Chan KS, Jakobsen KS, et al. Characterizing mixed microbial population dynamics using time-series analysis. ISME J 2008;2(7):707–15.

57. Becker SA, Feist AM, Mo ML, Hannum G, Palsson BO, Herrgard MJ. Quantitative prediction of cellular metabolism with constraint-based models: the COBRA Toolbox. Nat Protocols 2007;2(3):727–38.

CHAPTER 5

Computational Tools for Taxonomic Microbiome Profiling of Shotgun Metagenomes

Matthias Scholz, Adrian Tett and Nicola Segata

INTRODUCTION

Cultivation-free metagenomic studies of entire microbial communities (microbiomes) in the ocean, soil, and the human body have significantly improved our understanding of the role of the microbiome in natural environments and in human health and disease.[1–4] Sequencing the 16S ribosomal RNA (rRNA) gene is a popular cost-effective high-throughput technique[5] to assess the diversity of microbiomes, but shotgun sequencing of the whole genomic content of a community (the metagenome) provides a much richer snapshot of both the organismal composition and the metabolic potential of the community.[6] With the cost of high-throughput sequencing constantly decreasing, the number of whole-metagenome shotgun (WMS) datasets is rapidly increasing producing an unprecedented opportunity to unravel the composition, diversity, and function of these complex microbial communities.

Shotgun metagenomics produces extremely large datasets of short sequences that are very challenging to analyze. Recently, several computational approaches have been proposed to explore WMS data from different, albeit complementary, viewpoints (e.g., the taxonomic composition, the metabolic potential, and the phylogenetic diversity) but the potential of this technology is yet to be fully realized. Here we specifically focus on the task of taxonomic profiling of microbial communities from WMS samples, discussing the strategies developed so far and the ongoing challenges in the field.

Metagenomics for Microbiology. http://dx.doi.org/10.1016/B978-0-12-410472-3.00005-1

TAXONOMIC PROFILING WITH SHOTGUN METAGENOMICS

A WMS dataset[6] consists of millions of short sequence reads (genomic fragments) ideally representing all the microbes populating the sampled environment (Figure 5.1A and B). Identifying the organisms present in the microbial community and their abundances is usually the first step to unravel the biology of such communities and this task is referred to as *taxonomic profiling*. More formally, taxonomic profiling is the computational operation of inferring that taxonomic clades are populating a given microbial community and in what proportions (relative abundances).

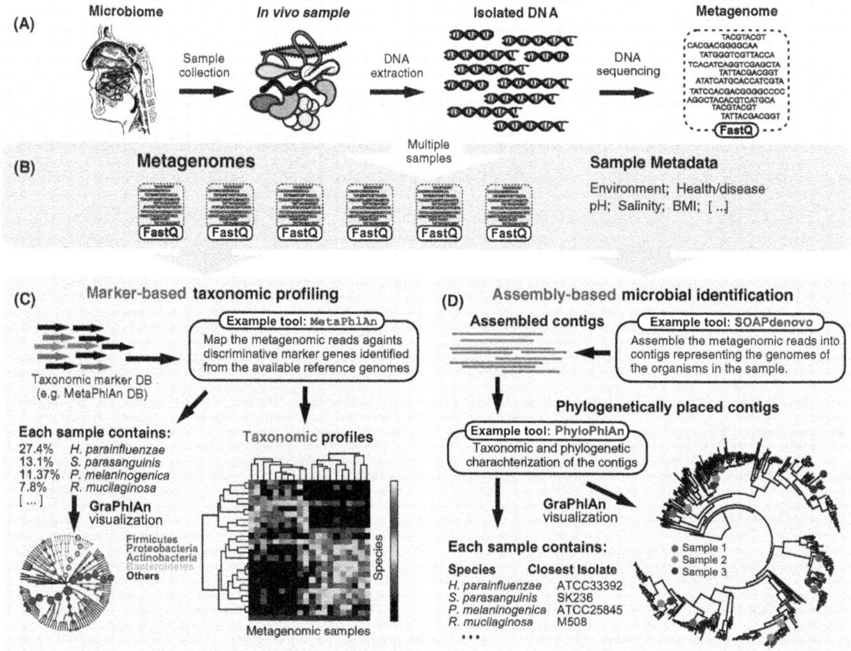

Fig. 5.1. A complete taxonomic profiling pipeline from sample collection to marker-based or assembly-based taxonomic profiling. (A) The DNA extracted and isolated from microbiome samples is sequenced by shotgun metagenomics that provides millions of short reads in each WMS sample. (B) A typical metagenomic dataset consists of sequenced samples in FastQ format (nucleotide sequence and corresponding quality scores[47]) and associated metadata. Using a (C) marker-based approach (e.g., MetaPhlAn[42]), short reads are mapped against representative and taxonomically informative genes. This results in species-level abundance profiles for each sample, which can then be merged and used for clustering or diversity analysis. (D) Short reads can alternatively first be assembled into longer contigs (e.g., with SOAPdenovo[16]) that are then placed into a phylogenetic tree (e.g., with PhyloPhlAn[23]) to explore the microbial diversity in each sample. Other profiling alternatives are discussed in the text, and a full step-by-step tutorial for the pipeline in panel (C) is available in the supplement of the article by Segata et al.[6]

A common feature of taxonomic profiling approaches is that they all, to varying degrees, rely on reference databases (of sequenced organisms) to assign taxonomic labels to the WMS sequences. This is necessary as no *a priori* taxonomic information is provided by the sequence itself. The sheer size of both the metagenome and the reference databases poses a major challenge to taxonomic profiling. For instance, single WMS samples from recent investigations[1–3] comprise between 10^9 and 10^{10} bases and the number of sequenced genomes in the public databases currently exceeds 15,000 (>40 Gb)[7,8] and rapidly increasing. Pioneering taxonomic profiling approaches were based on mapping with BLAST[9] and each sequencing read was searched against each microbial genome to ascribe taxonomy (usually by the best-hit policy),[10,11] but with current WMS and reference genome datasets this approach is unfeasible even with very large computational resources. Other challenges include biases in the reference databases that are largely populated by genomes of cultivable and well-characterized organisms.[7,8] Consequently, some portions of the tree of life are underrepresented. This lack of closely related reference genomes hampers the identification of phylogenetically conserved sequences as well as the assessment of the impact of horizontal gene transfer events. In addition, WMS data analysis faces the same common issues as other high-throughput next-generation sequencing projects including the relatively short read length (currently between 100 and 250 nucleotides) compared with traditional Sanger sequencing (~1000 nucleotides) and the non-negligible erroneous base call rates on the entire sequencing dataset.

Current approaches for WMS taxonomic profiling can be grouped based on how directly they make use of reference genomes. These are *Assembly*, *Compositional*, *Mapping*, and *Marker-based* approaches. Each of these approaches is discussed in the proceeding sections. Schemes of the two most popular approaches, Marker based and Assembly based are given in Figure 5.1C and D, respectively.

ASSEMBLY-BASED TAXONOMIC PROFILING OF MICROBIOME

To obtain a complete genomic snapshot of the environment sampled, the full-length genome sequence for each of the microbes present would need to be recovered. This is clearly idealistic, but by utilizing metagenomic *de novo* assembly techniques,[12] WMS reads can be first assembled into contigs, and at some instances, it is possible to reconstruct the genomes of the more dominant members of the community. After the initial assembly step, taxonomic or phylogenetic information is ascribed to each contig in a second step by sequence comparison to reference genomes. The scheme of assembly-based profiling is given in Figure 5.1D. For each of the two steps, several options are available and a few software packages combine both in a unique integrated pipeline such as MetAMOS,[13] MOCAT,[14] and Ray Meta.[15]

Single-genome assembly tools such as SOAPdenovo[16] have been directly applied on metagenomic data with varying degrees of success.[1,3] However, these tools are not optimized for metagenomic projects that consist of a mixture of genomes. Several metagenomic extensions have been developed to cope with metagenomic sequences including the computer memory issues because of the size of WMS samples, and the risk of generating trans-organismal chimeric contigs or scaffolds.[17] MetaVelvet[18] and Meta-IDBA[19] are two popular de Bruijn-based metagenomic assembly tools that work particularly well for the most abundant organisms in a WMS sample and high-quality assemblies have been obtained with *ad hoc* extensions of existing approaches.[20,21] The taxonomic placement of the reconstructed contigs is frequently performed by sequence mapping against reference genomes. Although this manual curation strategy can be effective in some cases, more precise and automatic tools have been specifically developed for this task including MetaPhyler[22] and PhyloPhlAn.[23] These tools, in addition to taxonomic assignment, also provide a phylogenomic assessment of the contigs by placing them in the context of the microbial tree of life (Figure 5.1D). In particular, PhyloPhlAn utilizes the 400 most conserved proteins within the sequenced members of the microbial phylogeny to infer the phylogenetic placement of new genomes or metagenomically assembled contigs. It has been shown[23] that even small contigs consisting of only 1% of the whole genome can be accurately rooted in the microbial phylogeny and taxonomic assignments can then be automatically inferred and manually inspected.

Assembly-based approaches are particularly suitable for microbiomes that include a large proportion of previously unseen (unsequenced) microbes. For these metagenomes, which are only partially covered by reference genomes, the advantage of assembly-based approaches is that they rely on a more indirect use of reference genomes compared with other profiling approaches that would miss the novel portions of the community. In contrast, for environments such as the human body for which extensive efforts have been made to sequence genomes that are representative of the overall microbial diversity,[24] assembly-free taxonomic profiling approaches are usually able to better capture the sequences for low-abundance members of the community that are particularly challenging to assemble. Currently, metagenomic assembly remains an active area of research as closely related organisms, highly conserved DNA regions, and horizontal gene transfer pose significant challenges in obtaining accurate assemblies.

COMPOSITIONAL APPROACHES FOR METAGENOMIC BINNING

Compositional approaches compare the intrinsic properties of sequences without being reliant on direct nucleotide or protein sequence alignment. Such intrinsic properties that are known to be good organismal signatures include variations in GC-content, codon usage bias, and the distribution of k-mers of variable length, with the latter being considered the most important compositional feature for comparison. In a compositional approach, the first step is to build a statistical model of species- or genus-specific intrinsic properties by preprocessing reference genomes (the so-called training step). The second step is applying this model to compare and classify the metagenomic reads. There are several different approaches to achieve these goals; for example, PhyloPythia/PhyloPythiaS[25] adopts a support vector machine classifier based on k-mer statistics. Different methods use other state-of-the-art machine-learning tools and these include Phymm[26] and NBC[27] that are based on Bayesian models and TACOA,[28] which adopts a k-nearest neighbor-based strategy.

Because compositional approaches avoid the computationally expensive sequence alignment, they usually permit quick running times. Similarly to assembly-based approaches, they have high generalizing capabilities showing good properties in classifying reads without closely related reference sequences. This capability is because of the fact that intrinsic sequence information is evolutionarily more conserved than nucleotide sequence homology. However, this ability comes at the expense of low discrimination power when closely related sequences are present in the reference databases. For this reason, compositional taxonomic profiling is usually limited to genus-level resolution. Moreover, the low discriminatory power is further exacerbated by very short sequencing reads. Combining compositional with mapping-based approaches can mitigate both shortcomings.

MAPPING-BASED RECRUITMENT OF METAGENOMIC READS

Mapping- or alignment-based methods categorize metagenomic reads based on sequence similarity with reference genomes. Currently, the most advanced tools are based on recent developments in DNA-based read-to-genomes mapping tools that in comparison to the first-generation BLAST-like tools are orders of magnitude faster, allowing millions of reads to be mapped against the human genomes in the order of a few minutes. They utilize compact indices (such as those based on the Burrows–Wheeler transform) to efficiently identify limited sets of subsequences of the reference genome on which the full alignment is performed. Although some profiling approaches still use BLASTN[9] as a mapping engine,[26,29] updating them to include these very fast algorithms such as Bowtie2,[30] SOAP2,[31] or BWA[32] is reasonably trivial. In some tools, the raw mapping is directly used as a proxy for the microbial community composition by naively assigning a taxonomic label to each metagenomic read based on the best hit. However, in the great majority of the cases, the raw output needs to be post-processed to resolve ambiguities in the mapping caused by conserved genomic regions, multiple reference genomes in the database from the same taxonomic clade, or reads mapping to the donor genome of horizontally transferred regions. These ambiguously assigned reads can be typically labeled using the lowest common ancestor approach that categorizes reads into the lowest possible taxonomic clade that includes all significant hits as is implemented in the web-based tools MG-RAST[33] and MEGAN.[34] More advanced phylogenetic-based tools are available[35] and hybrid approaches integrating compositional and mapping-based strategies have also been proposed including PhymmBL[26] that combines interpolated Markov models with sequence mapping, RITA[29] that implements a cascade of direct and translated mapping and Naïve Bayes compositional classifiers, and SPHINX[36] that limits the search space with tetranucleotides frequencies followed by translated mapping. These hybrids combine the computational performances and generalizability to high taxonomic level of compositional tools and the low taxonomic level (e.g., species) discriminability of mapping-based tools.

MARKER-BASED TAXONOMIC PROFILING

A large fraction of the genomic information available in reference genomes for the purposes of taxonomic profiling is at best uninformative (e.g., conserved sequences across multiple taxa) and occasionally even misleading (e.g., horizontally transferred genes). Marker-based approaches preprocess reference genomes to remove redundant and nondiscriminating sequences and focus on the most taxonomically informative markers (Figure 5.1C). As a consequence, this reduces the size of the reference genomes database and therefore decreases the computational requirements as the WMS samples are compared only against a fraction of each genome, the marker set. Two classes of markers have been exploited so far for taxonomic profiling: universal markers and clade-specific markers.

Universal markers are those sequences that: (i) are present in all microbes and (ii) possess variable regions that can be exploited as taxonomic or phylogenetic tags. The 16S ribosomal gene is the most notable example of universal marker that has been used for decades for taxonomic and phylogenetic investigation, and several cost-effective high-throughput sequencing approaches now target the diversity of a subset of its nine hypervariable regions.[5] The 16S rRNA gene is of course present in WMS samples as well (making up ~0.1% of the bacterial sequences) and it can, therefore, be used for taxonomic profiling with tools such as PhyloOTU.[37] To improve the robustness of the 16S rRNA taxonomic signal, additional highly conserved genes can be used (e.g., *hsp65* and *rpoB*[38]), and tools such as AMPHORA[39] and MetaPhyler[22] extend the set of universal markers further to several dozen including both bacteria and archaea genes thus improving the accuracy of the inferred community taxonomic profiles. PhyloPhlAn goes further by utilizing the 400 quasi-universal markers (i.e., present in almost all sequenced genomes) to infer the phylogenetic placement of organisms in the microbiome after a partial assembly step as detailed in the corresponding section above. Universal markers thus exploit few universally conserved genomic sequences that are expected to be present in yet-to-be-sequenced microbes, but cannot take advantage of nonubiquitous genes that constitute the majority of the microbial genomes.

The nonubiquitous regions of microbial genomes can be exploited focusing on clade-specific marker genes that are uniquely present in each taxonomic clade (e.g., each species). These genes are defined as core genes[40] within the given clade with no sequence similarity to any other gene outside the clade.[41] They are thus unique fingerprints of each microbial clade and can discriminate closely related organisms with high accuracy by just checking their presence (or absence) in the metagenome. MetaPhlAn[42] uses approximately 400,000 such clade-specific marker genes representing the whole tree-of-life to taxonomically characterize the organisms from WMS samples ensuring high accuracy, quantitative estimation, and subspecies resolution.[43] These tools make the most efficient use of the available reference genomes and therefore these approaches have the greatest potential for the development of fast and accurate metagenomic profiling.

PROFILING LARGE METAGENOMIC COHORTS: A CASE STUDY

Only a few large-scale metagenomic projects have utilized high-throughput shotgun sequencing to study the human microbiome with respect to health and disease. The two research efforts that generated the highest amount of shotgun metagenomics data for the human intestine are the MetaHIT project,[1] which focused on the gut microbiota of 124 individuals including 25 with inflammatory bowel diseases (~0.5 Tb), and a gut microbiota study of 345 Chinese type-2 diabetic patients and nondiabetic controls[2] (~1.5 Tb). In addition to the intestine, the Human Microbiome Project (HMP)[3] extended the study of the human microbiome to include 18 different body sites generating 3.5 Tb of sequences. With the cost of sequencing constantly decreasing, ongoing investigations will soon surpass these pioneering projects in terms of WMS data generated; thus, it is crucial to have taxonomic profiling tools able to accurately analyze WMS datasets in a computationally efficient manner.

As a case study, we present the taxonomic profiling of 656 WGS samples from the HMP dataset utilizing the analysis pipeline depicted in Figure 5.1C. MetaPhlAn[42] was applied to the 3.5 Tb of data, processing it at an average speed of 20,000 reads/second/CPU. This corresponds

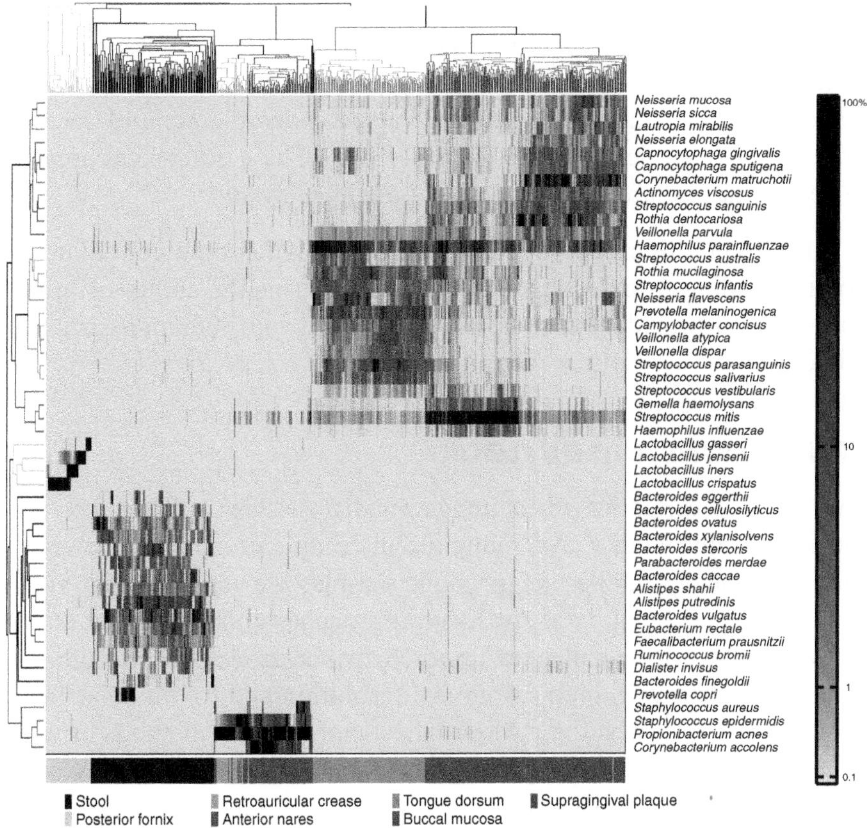

Fig. 5.2. Taxonomic profiles of human-associated shotgun sequencing microbiome samples from the Human Microbiome Project. In total, 656 metagenomic samples (Illumina 101 nt reads) from seven human body sites have been taxonomically characterized by MetaPhlAn (see Figure 5.1C) and the 50 most abundant species are shown. The samples are hierarchically clustered using the Bray–Curtis similarity and the species are grouped based on the correlation of their abundance patterns. The complete abundance matrix is available at http://www.hmpdacc.org/HMSMCP/.

to approximately 454 CPU hours, which is roughly equivalent to 2 days using 10 processors in parallel. In Figure 5.2, species-level taxonomic profiles are given for each of the 656 samples. Only the 50 most abundant species are shown in Figure 5.2 (complete profiles are available for download at http://www.hmpdacc.org/HMSMCP/). One of the findings is that the microbial-specific signature is stronger than the inter-patient variability for each body site. Only the microbiomes from the skin (retroauricular crease) and the anterior nares show a partially microbial overlapping. It is important to notice the inter-individual variability at the species level, an observation that would be missed if the profiling was limited to genus-level resolution. For example, several different *Bacteroides* species are common commensals of the gut, four different *Lactobacillus* species can dominate the vaginal microbiome, and several *Streptococcus* species colonize the oral cavity. These differences appear to be specific and are likely to be functionally relevant.[4]

Altogether, this case study highlights the role of computational taxonomic profiling from WMS samples and confirms the ability of available tools to provide a first taxonomic insight into human-associated microbiomes.

CONCLUSIONS AND OUTLOOK

Shotgun metagenomics is becoming an indispensable experimental tool for microbiome studies, and many recent computational developments to taxonomically profile metagenomic samples are exploiting the richness of the generated data. Available approaches utilize different strategies to reduce the challenges imposed by the size of WGS data, to generate profiles with high taxonomic resolution and to minimize false positives and false negatives. Although assembly-based methods usually have an advantage when exploring novel environments, they may miss low-abundance organisms in the microbiome, marker-based approaches preprocess the reference sequences to reduce their size and increase their discriminating power resulting in very fast and precise tools. Compositional and alignment-based methods attempt to give a comprehensive view by assigning taxonomies to as many reads as possible, but their strategies differ in terms of efficiency and taxonomic resolution.

Taxonomic profiling relies on well-characterized reference genomes. The profiling accuracy is thus deeply influenced by how well the available reference genomes database covers the biodiversity in the metagenomic sample under investigation. Human-associated microbiomes and the gut microbiome, in particular, are well represented in the reference genomes database because of their important role in human health and disease.[24] Novel approaches for genome sequencing and most notably single-cell sequencing is readdressing the bias and expanding the reference set for many diverse environments.[44] Metagenomic taxonomic profiling can thus take advantage from the richer set of genomes that is quickly being generated, although there are computational challenges in using larger reference sequence set and this will likely become the limiting factor.

Shotgun metagenomics also has the potential to identify non-bacterial members contributing to the biodiversity of microbial communities. For example, Achaea can be identified by all the tools discussed here whereas extensions for profiling the viral[45] and micro-eukaryotic (e.g., fungal[46,47]) diversity are currently only available in a few profiling methods. However, profiling of non-bacterial organisms has additional challenges because of the scarcity of available reference sequences and the lack of optimized DNA extraction protocols. Extending taxonomic profiling to include all domains of life, addressing issues of taxonomic resolution and computational efficiency are the most relevant challenges that novel computational methods for WMS analysis face, and so this field remains an active area of research. Effectively addressing these challenges will greatly contribute to better characterization and understanding of these complex microbial communities.

ACKNOWLEDGMENTS

This work has been supported by the People Programme (Marie Curie Actions) of the European Union's Seventh Framework Programme (FP7/2007-2013) under REA grant agreement no. PCIG13-GA-2013-618833, startup funds from the Centre for Integrative Biology (University of Trento), by MIUR "Futuro in Ricerca" E68C13000500001, and from Fondazione Caritro to the co-author of this chapter, Nicola Segata.

REFERENCES

1. Qin J, Li R, Raes J, Arumugam M, Burgdorf KS, Manichanh C, et al. A human gut microbial gene catalogue established by metagenomic sequencing. Nature 2010;464(7285):59–65.
2. Qin J, Li Y, Cai Z, Li S, Zhu J, Zhang F, et al. A metagenome-wide association study of gut microbiota in type 2 diabetes. Nature 2012;490(7418):55–60.
3. The Human Microbiome Project Consortium. A framework for human microbiome research. Nature 2012;486(7402):215–21.
4. Huttenhower C, Gevers D, Knight R. Structure, function and diversity of the healthy human microbiome. Nature 2012;486(7402):207–14.
5. Hamady M, Knight R. Microbial community profiling for human microbiome projects: Tools, techniques, and challenges. Genome Res 2009;19(7):1141–52.

6. Segata N, Boernigen D, Tickle TL, Morgan XC, Garrett WS, Huttenhower C. Computational meta'omics for microbial community studies. Mol Syst Biol 2013;9:666.
7. Sayers EW, Barrett T, Benson DA, Bryant SH, Canese K, Chetvernin V, et al. Database resources of the National Center for Biotechnology Information. Nucleic Acids Res 2009;37(Database issue):D5–D15.
8. Markowitz VM, Chen IM, Palaniappan K, Chu K, Szeto E, Grechkin Y, et al. IMG: the Integrated Microbial Genomes database and comparative analysis system. Nucleic Acids Res 2012;40(Database issue):D115–22.
9. Altschul SF, Gish W, Miller W, Myers EW, Lipman DJ. Basic local alignment search tool. J Mol Biol 1990;215(3):403–10.
10. Rusch DB, Halpern AL, Sutton G, Heidelberg KB, Williamson S, Yooseph S, et al. The Sorcerer II Global Ocean Sampling expedition: northwest Atlantic through eastern tropical Pacific. PLoS Biol 2007;5(3):e77.
11. Shi Y, Tyson GW, Eppley JM, DeLong EF. Integrated metatranscriptomic and metagenomic analyses of stratified microbial assemblages in the open ocean. ISME J 2011;5(6):999–1013.
12. Nagarajan N, Pop M. Sequence assembly demystified. Nat Rev Genet 2013;14(3):157–67.
13. Treangen TJ, Koren S, Sommer DD, Liu B, Astrovskaya I, Ondov B, Darling AE, Phillippy AM, Pop M. MetAMOS: a modular and open source metagenomic assembly and analysis pipeline. Genome Biol 2013;14(1):R2.
14. Kultima JR, Sunagawa S, Li J, Chen W, Chen H, Mende DR, et al. MOCAT: a metagenomics assembly and gene prediction toolkit. PloS One 2012;7(10):e47656.
15. Boisvert S, Raymond F, Godzaridis É, Laviolette F, Corbeil J. Ray Meta: scalable de novo metagenome assembly and profiling. Genome Biol 2012;13(12):R122.
16. Li R, Zhu H, Ruan J, Qian W, Fang X, Shi Z, et al. De novo assembly of human genomes with massively parallel short read sequencing. Genome Res 2010;20(2):265–72.
17. Koren S, Treangen TJ, Pop M. Bambus 2: scaffolding metagenomes. Bioinformatics 2011; 27(21):2964–71.
18. Namiki T, Hachiya T, Tanaka H, Sakakibara Y. MetaVelvet: an extension of Velvet assembler to de novo metagenome assembly from short sequence reads. Nucleic Acids Res 2012;40(20):e155.
19. Peng Y, Leung HC, Yiu SM, Chin FY. Meta-IDBA: a de Novo assembler for metagenomic data. Bioinformatics 2011;27(13):i94–i101.
20. Albertsen M, Hugenholtz P, Skarshewski A, Nielsen KL, Tyson GW, Nielsen PH. Genome sequences of rare, uncultured bacteria obtained by differential coverage binning of multiple metagenomes. Nat Biotechnol 2013;31(6):533–8.
21. Hess M, Sczyrba A, Egan R, Kim TW, Chokhawala H, Schroth G, et al. Metagenomic discovery of biomass-degrading genes and genomes from cow rumen. Science 2011;331(6016):463–7.
22. Liu B, Gibbons T, Ghodsi M, Treangen T, Pop M. Accurate and fast estimation of taxonomic profiles from metagenomic shotgun sequences. BMC Genomics 2011;12(Suppl 2):S4.
23. Segata N, Bornigen D, Morgan XC, Huttenhower C. PhyloPhlAn is a new method for improved phylogenetic and taxonomic placement of microbes. Nat Commun 2013;4:2304.
24. Nelson KE, Weinstock GM, Highlander SK, Worley KC, Creasy HH, Wortman JR, et al. A catalog of reference genomes from the human microbiome. Science 2010;328(5981):994–9.
25. McHardy AC, Martin HG, Tsirigos A, Hugenholtz P, Rigoutsos I. Accurate phylogenetic classification of variable-length DNA fragments. Nat Methods 2007;4(1):63–72.

26. Brady A, Salzberg SL. Phymm and PhymmBL: metagenomic phylogenetic classification with interpolated Markov models. Nat Methods 2009;6(9):673–6.

27. Rosen GL, Reichenberger ER, Rosenfeld AM. NBC: the Naive Bayes Classification tool web-server for taxonomic classification of metagenomic reads. Bioinformatics 2011;27(1):127–9.

28. Diaz NN, Krause L, Goesmann A, Niehaus K, Nattkemper TW. TACOA: taxonomic classification of environmental genomic fragments using a kernelized nearest neighbor approach. BMC Bioinform 2009;10:56.

29. MacDonald NJ, Parks DH, Beiko RG. Rapid identification of high-confidence taxonomic assignments for metagenomic data. Nucleic Acids Res 2012;40(14):e111.

30. Langmead B, Salzberg SL. Fast gapped-read alignment with Bowtie 2. Nat Methods 2012; 9(4):357–9.

31. Li R, Yu C, Li Y, Lam TW, Yiu SM, Kristiansen K, Wang J. SOAP2: an improved ultrafast tool for short read alignment. Bioinformatics 2009;25(15):1966–7.

32. Li H, Durbin R. Fast and accurate short read alignment with Burrows-Wheeler transform. Bioinformatics 2009;25(14):1754–60.

33. Meyer F, Paarmann D, D'Souza M, Olson R, Glass EM, Kubal M, et al. The metagenomics RAST server – a public resource for the automatic phylogenetic and functional analysis of metagenomes. BMC Bioinform 2008;9:386.

34. Huson DH, Auch AF, Qi J, Schuster SC. MEGAN analysis of metagenomic data. Genome Res 2007;17(3):377–86.

35. Berger SA, Stamatakis A. Aligning short reads to reference alignments and trees. Bioinformatics 2011;27(15):2068–75.

36. Mohammed MH, Ghosh TS, Singh NK, Mande SS. SPHINX – an algorithm for taxonomic binning of metagenomic sequences. Bioinformatics 2011;27(1):22–30.

37. Sharpton TJ, Riesenfeld SJ, Kembel SW, Ladau J, O'Dwyer JP, Green JL, Eisen JA, Pollard KS. PhylOTU: a high-throughput procedure quantifies microbial community diversity and resolves novel taxa from metagenomic data. PLoS Computat Biol 2011;7(1):e1001061.

38. Ludwig W, Klenk H-P. Overview: A phylogenetic backbone and taxonomic framework for procaryotic systematics. In: Boone DR, Castenholz RW, Garrity GM editors. Bergey's Manual of Systematic Bacteriology. New York, NY: Springer-Verlag; 2000. pp. 49–65.

39. Wu M, Scott AJ. Phylogenomic analysis of bacterial and archaeal sequences with AMPHORA2. Bioinformatics 2012;28(7):1033–4.

40. Segata N, Huttenhower C. Toward an efficient method of identifying core genes for evolutionary and functional microbial phylogenies. PloS One 2011;6(9):e24704.

41. Huang K, Brady A, Mahurkar A, White O, Gevers D, Huttenhower C, Segata N. MetaRef: a pan-genomic database for comparative and community microbial genomics. Nucleic Acids Res 2014;42(D1):D617–24.

42. Segata N, Waldron L, Ballarini A, Narasimhan V, Jousson O, Huttenhower C. Metagenomic microbial community profiling using unique clade-specific marker genes. Nat Methods 2012;9(8):811–4.

43. Scher JU, Sczesnak A, Longman RS, Segata N, Ubeda C, Bielski C, et al. Expansion of intestinal Prevotella copri correlates with enhanced susceptibility to arthritis. eLife 2013;2:e01202.

44. Rinke C, Schwientek P, Sczyrba A, Ivanova NN, Anderson IJ, Cheng JF, et al. Insights into the phylogeny and coding potential of microbial dark matter. Nature 2013;499(7459):431–7.

45. Culley AI, Lang AS, Suttle CA. Metagenomic analysis of coastal RNA virus communities. Science 2006;312(5781):1795–8.

46. Findley K, Oh J, Yang J, Conlan S, Deming C, Meyer JA, et al. Topographic diversity of fungal and bacterial communities in human skin. Nature 2013;498(7454):367–70.

47. Cock PJ, Fields CJ, Goto N, Heuer ML, Rice PM. The Sanger FASTQ file format for sequences with quality scores, and the Solexa/Illumina FASTQ variants. Nucleic Acids Res 2010;38(6):1767–71.

CHAPTER 6

Hypothesis Testing of Metagenomic Data

Patricio S. La Rosa, Yanjiao Zhou, Erica Sodergren, George Weinstock and William D. Shannon

INTRODUCTION

Statistical hypothesis testing is well established in clinical trials and basic research for objectively deciding whether data from multiple groups come from the same or different distributions.[1,2] The use of formal statistical hypothesis testing will become more important in human microbiome research[3–5] as this field moves from technical development and basic science discovery to translational medicine and environmental studies. To statistically test for differences across groups, a formal set of logical steps is always followed, which we define here for the general case. Once these steps are outlined, we will illustrate how they are applied to the microbiome data.

Formally, hypothesis testing is defined as a statistical procedure to decide whether the data collected provides enough evidence to accept a *null hypothesis* (e.g., the data from groups that come from the same distribution) or to reject it in favor of an *alternative hypothesis* (e.g., they come from different distributions).[1,2] For example, an investigator might be interested in the comparison of the average heights of jockeys and basketball players whose (hypothetical) distributions are shown in Figure 6.1. The histogram shows that the heights for these two populations are both (visually) normally distributed and that, as expected, the heights of the basketball players are more than those of the jockeys. To decide if these two distributions are the same or different, the statistician formulates the problem into a hypothesis, decides on the statistical test (a t-test in this simple example), and applies a formula to calculate the P value.[1]

To design an experiment, an investigator determines how many samples are needed to have a specified level of power to correctly conclude that the groups are different. To do this, four questions must be

Metagenomics for Microbiology. http://dx.doi.org/10.1016/B978-0-12-410472-3.00006-3

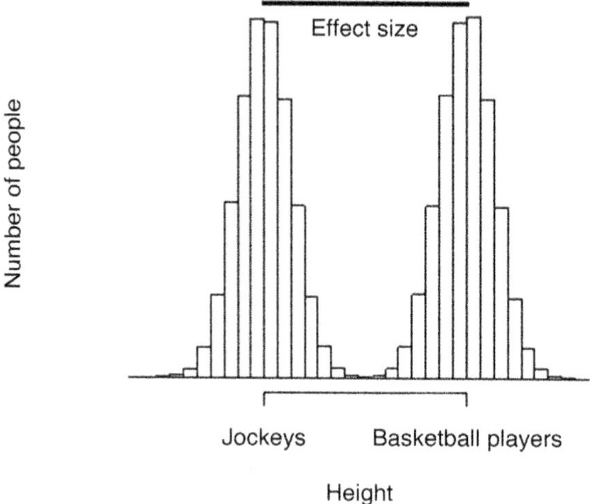

Fig. 6.1. Example of histograms of the height distributions from Jockeys and basketball players.

answered. First, what are the distributions that generate the data? In the example above, we believe that the heights of jockeys and basketball players are normally distributed with means and variances whose values are available from prior data. Second, how big of an *effect size* is being tested? As indicated in Figure 6.1, the distance between the two means is denoted as the *effect size* of the hypothesis,[2,6] and the further apart these two means are the larger the effect size is. It should be noted that the definition of an effect size is specific to the type of data being analyzed. In the above case, it is simply the difference between the two means. A more complicated effect size, for example, is the hazard ratio comparing survival curves using a Cox proportional hazards model.[1] Attention to how the effect size is defined and measured is important. Third, what is the test statistic that will be used to reject or not reject the null hypothesis? The above example clearly requires the use of the t-test,[1] but issues such as symmetric distributions with wide tails would indicate that a Wilcoxon test[1] would be more appropriate if the heights were not normally distributed. Fourth, what level of statistical performance is required? Generally accepted performance is $P < 0.05$ for significance with 80% power.[6] These values can be changed that have direct impact on the sample size, as we will indicate in the text below. For example, the analyst may consider to use a lower level of significance (e.g., $P < 0.01$) when it is important to avoid incorrectly rejecting the null hypothesis, or to use a larger power (e.g., 90%) when it is important to avoid not rejecting the null hypothesis when it should be rejected. Setting these parameters are project specific and should involve discussion among the key researchers designing the experiment including the primary investigator and co-investigators, funding agency, and biostatistician.

The *P* value, power, effect size, and sample size are all completely interrelated.[6] If the *P* value and power are held constant, a larger (smaller) effect size results in a smaller (larger) sample size. Given that the *P* value and effect size are held constant, increasing (decreasing) the sample size increases (decreases) the power of the study. This indicates why it is important to specify all four of the parameters when designing an experiment.

The recipe presented here is repeated for any formal statistical hypothesis test and design of an experiment for one, two, or multiple groups, whether univariate, multivariate, multiple univariate tests with adjustments for multiple testing, time series, or survival analyses. In the remaining of this chapter, we present three examples of hypothesis testing for microbiome data: (1) compare the diversity measure of bacterial species present across groups; (2) compare the frequency of a taxon of interest across groups; and (3) compare the frequency of taxa across groups.

METAGENOMIC DATA

In this chapter, the examples are comparisons of taxonomic-labeled metagenomic data as formatted in Table 6.1. The entries in the table are the number of sequence reads assigned to a particular taxon by sample designated X_{ik} indicating the number of taxon "*k*" in subject "*i.*". The

Table 6.1 Basic Data Structure and Format of Metagenomic Samples to Perform Hypothesis Testing and Statistical Analysis

	Taxa				
Sample	**1**	**2**	**...**	**K**	**#Reads/Sample**
1	X_{11}	X_{12}	...	X_{1K}	$X_{1\bullet}$
2	X_{21}	X_{22}	...	X_{2K}	$X_{2\bullet}$
⋮	⋮	⋮	⋱	⋮	⋮
N	X_{N1}	X_{N2}	...	X_{NK}	$X_{N\bullet}$
#Reads/Taxon	$X_{\bullet 1}$	$X_{\bullet 2}$	...	$X_{\bullet K}$	

total number of reads for each sample is denoted by the symbol X_{i*} and the total number of reads per taxon is denoted by the symbol X_{*k}. While this notation may seem clumsy, it is a standard format used in statistics to describe data and calculations.

In experiments with multiple groups, the rows will be divided by group. For example, if a two-group study was done, the first 10 rows might be samples from group 1 and the second 10 rows from group 2. In any case, the basic data structure is still as defined in Table 6.1.

One additional consideration involves the issue of standardization or normalization of the data. Suppose we wish to analyze percentages of each taxon within the samples. Algebraically, this involves dividing the reads in each row by the total number of reads for that row, and multiplying this by 100. For example, the percent of taxon 1 in sample 1 is calculated as $(X_{11}/X_{1*}) \times 100$. Since different investigators chose to transform their data differently (see, e.g., Legendre and Legendre,[7] and references therein), we will not provide an exhaustive list of formula here, but do encourage the reader to carefully think about how they are changing their data. In the Dirichlet-multinomial (DM) model (presented below),[8] raw count data is used with no need for rarefaction (a popular data transformation among ecologist used to compare diversity indexes) that introduces loss of information and has been shown to increase false-positive errors.[9] The investigator should keep in mind that transforming data should be done based on solid theoretical foundation and for a specific purpose such as variance stabilization.[10]

COMPARE DIVERSITY ACROSS GROUPS

Our first microbiome example involves comparison of species diversity across groups. The number and variety of individual taxa present in a metagenomic sample can be summarized using an index measure (i.e., single number), such as the Shannon diversity index[7] ($H_i = \sum_{i=1}^{K} (X_{ik}/X_{i*})\log(X_{ik}/X_{i*})$) or species evenness[7] ($J_i = H_i/\log(K)$ *where Ki is the number of species present in subject i*), calculated for each sample. When an investigator is interested in seeing if there is more diversity in one group versus another, these measures can form the basis of the hypothesis test, power, and sample size calculations. Depending on how these values are distributed and the number of groups, a standard *t*-test, analysis of variance, or corresponding nonparametric test can be used.[1]

To illustrate, we compare the Shannon diversity calculated on Vervet monkeys from a study comparing the impact on the stool microbiome of a typical American diet (TAD) versus a commercially available monkey feed (CHOW). For each of the 136 monkeys, the microbiome data, formatted as represented in Table 6.1, for each sample, was used to calculate the Shannon diversity for that monkey. The histograms in Figure 6.2 shows the percent of the 136 monkeys (*Y* axis) binned into Shannon diversity values (*X* axis) divided into two groups. Figure 6.2 shows that TAD results in higher diversity since the histogram for this group is shifted to the right (higher diversity values) relative to the CHOW diet group. To test the null hypothesis of no difference in diversity,

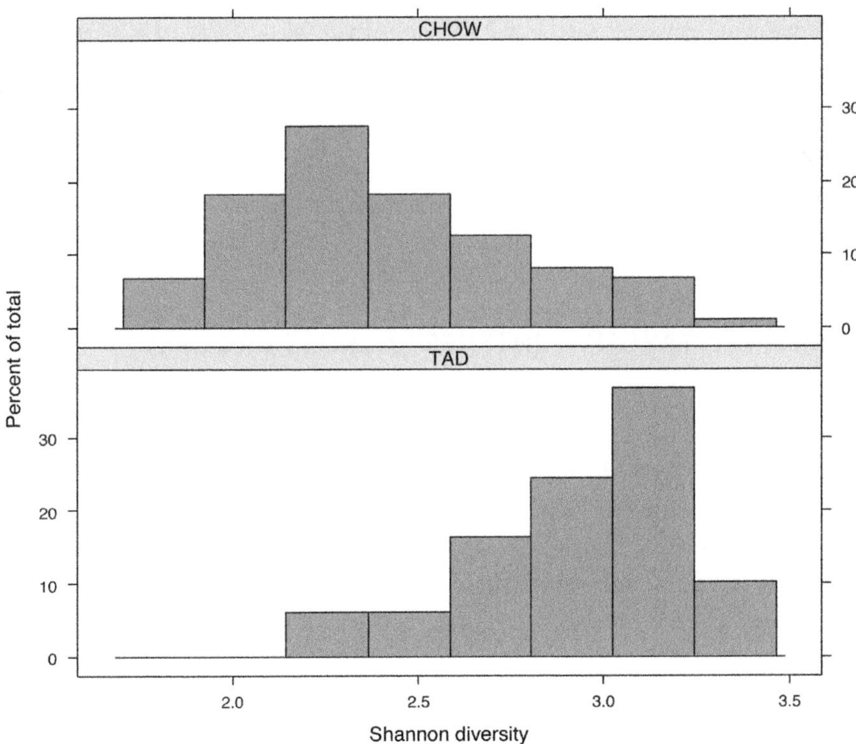

Fig. 6.2. Histograms of the Shannon diversity distributions computed on taxa composition data from the stool of a group of Vervet monkeys feeding Chow (upper panel) and another group feeding on typical american diet (TAD) (lower panel).

a t-test was used resulting in $P < 0.0001$ ($t = 9.4$, $df = 118.4$) leading us to reject the null hypothesis of no difference in favor of the alternative that the Shannon diversities are different in the two groups. Note that the distributions are slightly skewed, so a Wilcoxon test was done giving $P < 0.0001$ leading to the same conclusion. Alternatively, a transformation to induce normality could be used (i.e., Box-Cox transformation[7]), although in this case, little change in the P value would result since the t-test is robust to non-normality.[1]

In experimental design, sample sizes are calculated to ensure adequate power to detect a difference in groups based on pilot data. In this data, the mean and the standard deviation for the TAD group are 2.94 and 0.288, respectively, and for the CHOW group are 2.41 and 0.359, respectively. From this data, the effect size is simply the difference between the means (delta = 2.94 − 2.41), and calculating the power or sample size can be done in almost any statistical software package; for example, in R, we used power.t.test(n = 2:20, delta = 2.94 − 2.41, sd = sqrt(0.1)) to generate the data for Figure 6.3.

From this power calculation we see that a sample size of seven Vervet monkeys per group, randomly assigned to either TAD or CHOW, will result in 80% power to reject the null hypothesis than the Shannon diversity is the same in the two groups.

COMPARE A TAXON OF INTEREST ACROSS GROUPS

Consider now the problem of testing a single taxon specified *a priori* across groups. It is important to specify the taxon *a priori* to ensure that the researcher is not looking at the data to see which taxa seem most different, then running tests to prove that what they saw is real. Statistically, testing data that has been looked at and selected based on apparent differences will inflate the false-positive rate (i.e., rejects the null hypothesis when it should not be rejected too often).

When the investigator wants to compare the abundance of the taxon across groups, it is important to standardize the data to a common scale. For example, converting the abundance to the percentage of reads in the sample (by dividing the taxon count by the total number of reads times 100) scales the data to "the number of taxon per 100 reads." Once

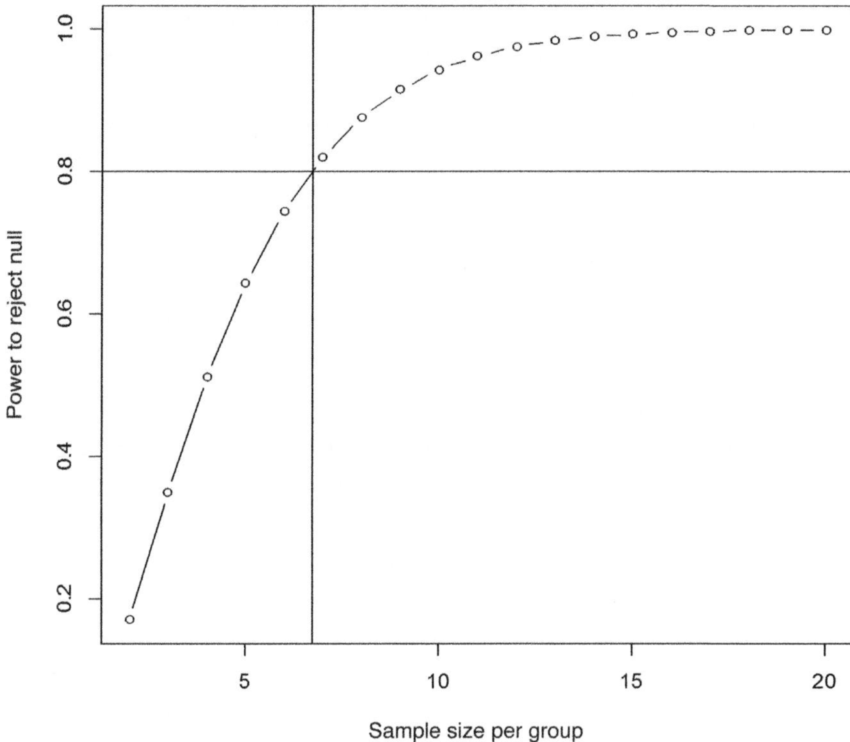

Fig. 6.3. Power of the t-test as a function of the sample size per group for an effect size equal to 0.53, the difference between the average Shannon diversity index of the TAD and the Chow group.

the taxon count is rescaled to a common value (e.g., percentage), it can be compared across groups in a similar way as was done for diversity measures. Note that problems may arise if a large proportion of the percentages are between 0–20% and 80–100%, in which case an inverse sine transformation should be considered for variance stabilization.[10] More complex modeling of the count data, such as negative-binomial regression, might also be considered to see if taxon count is impacted by subject phenotype such as age, gender, and health status, but this is beyond the scope of this chapter.

To illustrate a second test of a single *a priori* specified taxon, consider the case where the investigator wants to compare whether the taxon is present at different rates across groups. The count data in Table 6.1 for a taxon would be transformed to 0 if the taxon is absent in the sample or to 1 if the taxon is present in the sample. With categorical data such as this, a chi-square test is used. To illustrate, we compared the rate that *Streptococcus* appears in samples from the left retroauricular crease and the gastrointestinal tract from the Human Microbiome Project (HMP) data. Table 6.2 shows the distribution of these rates – 167 (92%) out of 181 left-retroauricular crease samples had *Streptococcus*, while 135 (65%) out of 209 stool samples did. To test the null hypothesis that of no difference in the rates of occurrence across the groups, a chi-square test gave a P value < 0.0001 ($X^2 = 40.9$, $df = 2$) leading us to reject the null hypothesis of no difference in favor of the alternative that the groups do have different rates of occurrence.

If an investigator is interested in confirming these results in a second experiment, the sample size (needed to ensure adequate power) can be computed using Figure 6.4 as pilot data. To compute power for this example, we first estimate the effect size. In the case of Shannon diversity above, the effect size was easily calculated as the difference in means. With categorical data, we do not use the difference in rates but require a different measure of effect size, such as Cramer's Phi. Other measures of effect size such as the odds ratio or relative risk that might be of interest and selection from these alternatives is based on the goals of the investigator. Using the pilot data in Table 6.2, Cramer's Phi = 0.324, and setting the significance level at 0.05, a single function call in R (pwr.chisq.test(w = 0.324, df = 1, N = 1:200, sig.level = 0.05)) gives the power for various sample sizes as shown in Figure 6.4. From this, it is determined that 31 samples are needed in each group to correctly reject the null hypothesis with 80% power. For 90% power, approximately 45 samples per group are needed.

A comment about multiple testing – in practice, many investigators will compare each taxon across groups separately resulting in a multiple testing problem.[2,11] To understand why this needs adjustment, it is necessary to understand what the P value means. If in a testing

Table 6.2 Distribution of the *Streptococcus* Rate Across Stool and Left-Retroauricular Crease Samples Obtained from the Human Microbiome Project

Body Site	Presence	Absence	Total
Left crease	167 (92%)	14 (8%)	181
Stool	135 (65%)	74 (35%)	209

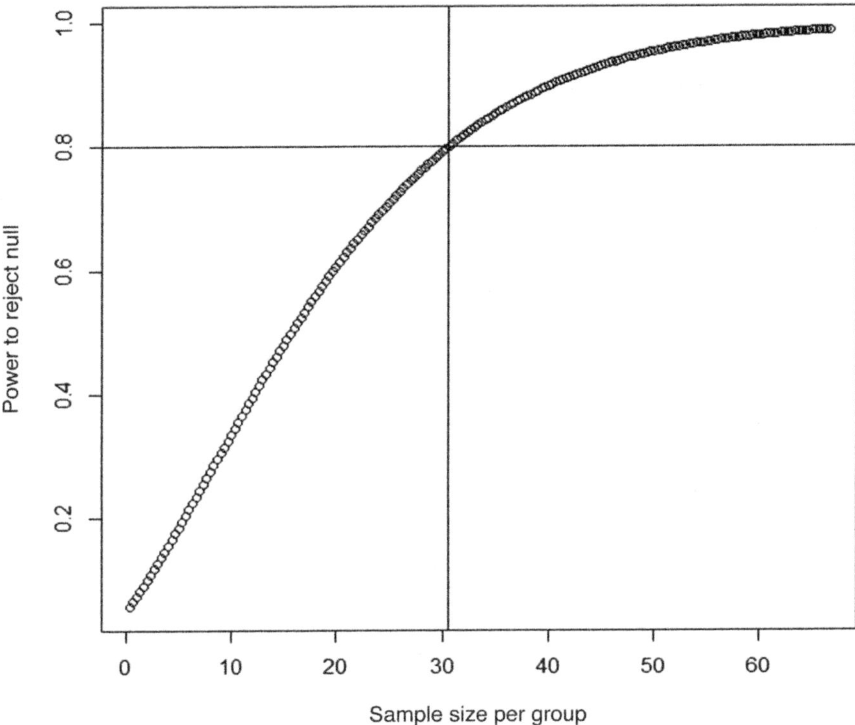

Fig. 6.4. Power of the chi-squared test as a function of the sample size per group for an effect size given by Cramer's-Phi = 0.354, a measure of the difference between the rate of Streptococcus found in metagenomic samples from the left retroauricular crease and the gastrointestinal tract populations from the human microbiome project.

situation the null hypothesis is true (i.e., the distributions of the taxon across groups is the same), the P value tells us the probability that in our data, the observed difference across groups occurred strictly because of chance. When we accept $P \leq 0.05$, we are explicitly accepting that we are OK with a 5% chance of saying that the groups are different when they are not (i.e., known as a Type I error, rejecting the null hypothesis when the null hypothesis is true, or a false positive).

Consider now the case where two taxa are separately tested and we use a $P \leq 0.05$ as the level of significance for each test. We might incorrectly conclude that taxon A is different across groups when it is not, or incorrectly conclude that taxon B is different across groups when it is not. However, we must also take into account the possibility that we incorrectly conclude that both A and B are different across groups when they are not. The true Type I error in this case is easily calculated as $1 - (1 - 0.05)^2 = 0.0975 >> 0.05$. So when the investigator thinks that their chance of committing a Type I error (making a false-positive conclusion) is 5%, it is in fact 9.75%. For 10 separate comparisons $(1 - (1 - 0.05)^{10} = 0.4013)$, this increases to a 40% chance of committing a Type I error.

Because of space limitations, we cannot fully explore this topic here but want the reader to understand the importance of protecting against this inflation in Type I error.[12]

COMPARE THE FREQUENCY OF ALL TAXA ACROSS GROUPS

Although an investigator could approach the comparison of multiple taxa by comparing the abundances of each taxon across groups separately adjusting for multiple comparisons as described above, this approach is generally less powerful than multivariate approaches as it does not take into account the interactions that exist between taxa. Multivariate statistical methods were invented for exactly this type of problem.

A multivariate distribution that applies to metagenomic data, taking into accounts the interactions or correlations among the taxa, is the Dirichlet-Multinomial (DM) model of taxa counts.[8,13,14] Parametric models, such as this, improve the analysis of data compared with nonparametric approaches and generally simplify calculation of *P* values, sample sizes, power calculations, measures of error, and confidence intervals. In addition, natural measures of effect size often are available from the parameters.

To illustrate, we present two analyses. The first compares the average taxa frequency across metagenomic samples from saliva and throat and formally tests the null hypothesis that these two microbiome populations are the same. Figure 6.5 shows the taxa frequency for the HMP samples for both groups at the order level. For example, *Lactobacillales* has an average abundance of approximately 40% in throat samples and approximately 18% in saliva samples. Using the test formula derived for the DM distribution, we calculated $P = 0.038$ ($X_{mc} = 78.33$, $df = 11$), indicating that the null hypothesis is rejected, and we conclude that the two groups have different proportions of these taxa.

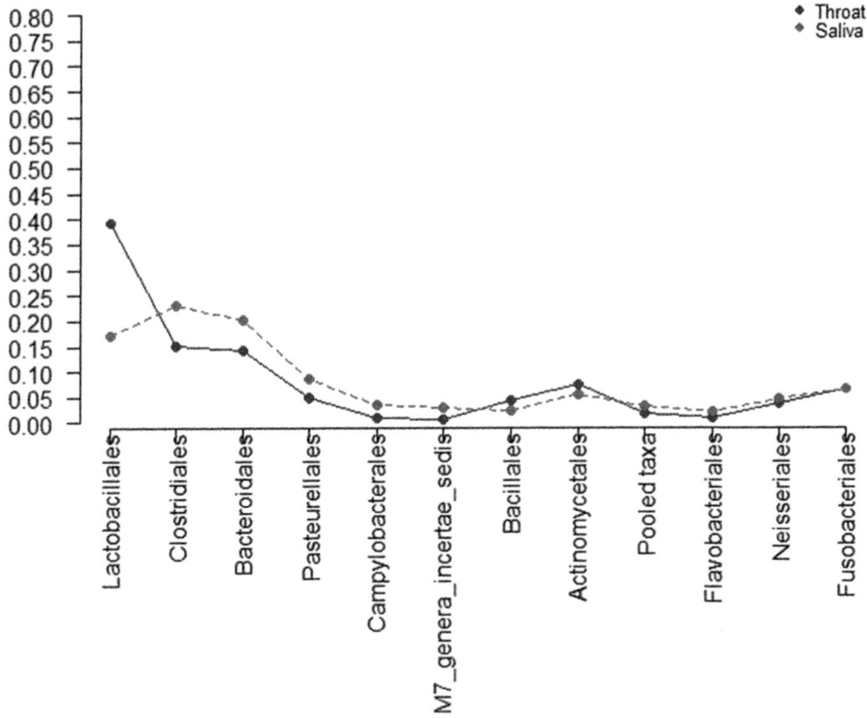

Fig. 6.5. Average taxa frequency computed across metagenomic samples from the saliva and throat populations obtained from the human microbiome project.

If an investigator wanted to confirm these results in a second experiment, the sample size (needed to ensure adequate power) can be computed using Figure 6.5 as pilot data. To compute power for this example, we first need to compute the effect size. As mentioned previously, the definition of effect size varies for different types of data. For the Shannon diversity example, the effect size was the difference in means, while for the categorical data using chi-square statistics, it was a more complicated calculation (i.e., Cramer's Phi). For this problem, the effect size is, intuitively, a measure of the distance between the two lines in Figure 6.5, and is calculated as a modified Cramer's Phi measure.[8] The further apart the lines are the larger the effect size is, and as in all power calculations, the fewer samples that would be needed for a given level of power.

Performing power calculations for metagenomics data requires the R-package HMP: hypothesis testing and power calculations for comparing metagenomic samples[15] developed by authors of this chapter.

To calculate power, the investigator estimates the DM parameters for each group (part of the R package), computes the effect size, sets the significance level (e.g., 5%), and specifies the number sequences that will be generated on average for each sample. In practice, we suggest that a range of sample sizes and depth of sequencing be specified to generate a table of power values as shown in Table 6.3. To compare the curves in Figure 6.5, the investigator should sequence seven samples per group to guarantee 90% power to correctly reject the null hypothesis. While power increases only slightly with more reads, it is often justified to do deep sequencing to increase the probability of finding rare taxa; however, this is outside the discussion of this chapter.

In the above example, the distributions were far apart (i.e., the lines in Figure 6.5 were far apart) with a calculated effect size equal to 0.27. In a second example, we use subgingival and supragingival plaque samples to show the impact of effect size on sample size and power. In this example, the effect size = 0.07 is a much smaller difference that is confirmed visually in Figure 6.6 where the lines are closer together. Table 6.4 shows the power analysis for this data using 1% and 5% significance levels where we see that about 25 samples per group will be needed to have approximately 90% power. Also note that as the significance level gets smaller (i.e., $P \leq 0.01$ to reject the null hypothesis), the power decreases for the same number of samples.

A comment about using the wrong distribution – using an incorrect statistical model for data can often lead to incorrect results, and

Table 6.3 Power Calculation as a Function of Number of Sequence Reads and Sample Size for the Comparison of the Average Taxa Frequency from the Throat and Saliva Populations Obtained from the Human Microbiome Project, Using 5% Significance Levels

Power	Dirichlet-Multinomial Number of Sequence Reads		
Number of Subjects	**1000**	**5000**	**10,000**
2	34.3%	37.9%	38.2%
5	75.8%	76.6%	77.2%
6	85%	86.6.%	87.1%
7	92.1%	92.7%	92.9%
8	96.1%	96.3%	96.5%

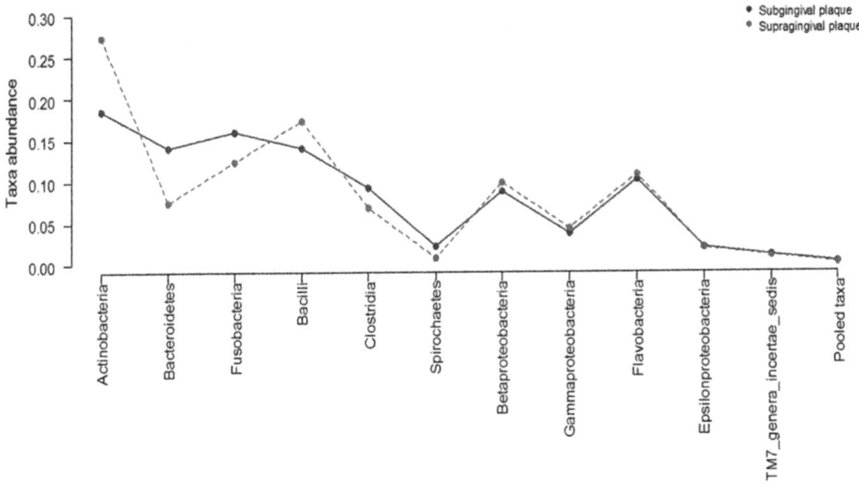

Fig. 6.6. Average taxa frequency computed across metagenomic samples from the subgingival and supragingival plaque populations obtained from the human microbiome project.

therefore, care must be taken when designing a study. A natural first pass look at metagenomics count data might lead a statistician to consider using a multinomial model to perform hypothesis testing, as well as power and sample size calculations. We and others have shown that the multinomial model is incorrect since it cannot capture the excess variability present in metagenomic data (technically, this is defined as overdispersion). Ignoring the overdispersion leads to designing studies with significantly lower power than the investigator believes. For example, Table 6.5 shows the power estimated using the wrong multinomial

Table 6.4 Power for Comparison of the Average Taxa Frequency from the Subgingiva and Supragingiva HMP Samples at 1% and 5% Significance Levels

	Alpha = 1%			Alpha = 5%		
Power	**Number of Sequence Reads**			**Number of Sequence Reads**		
Number of Subjects	**1000**	**5000**	**10,000**	**1000**	**5000**	**10,000**
10	29.45%	29.83%	29.89%	52.79%	52.91%	53.20%
15	55.26%	56.16%	56.16%	77.10%	77.88%	77.98%
25	89.44%	90.03%	90.00%	96.80%	97.02%	97.13%
50	99.96%	99.98%	99.96%	99.99%	99.99%	99.99%

Table 6.5 Comparison Between the Power of a Test Statistics Based on Multinomal and DM Models for Throat and Saliva HMP Samples Using 5% Significance Level

Power	Multinomial Number of Sequence Reads					Dirichlet-Multinomial Number of Sequence Reads		
Number of Subjects	50	100	1000	2000	5000	1000	5000	10,000
2	>73.3%	>96.7%	>99.9%	>99.9%	>99.9%	34.3%	37.9%	38.2%
5	>99.9%	>99.9%	>99.9%	>99.9%	>99.9%	75.8%	76.6%	77.2%

distribution for comparing the saliva and throat data. This indicates two samples per group and 1000 sequences is sufficient to have >99.9% power, when in reality, using the correct DM model, this sample size has only 34.3% power. Using the wrong distribution will result in significant increase in deciding that the groups are not different when they are (Type II error or false negative).

DISCUSSION AND FUTURE RESEARCH

This chapter has introduced power and sample size calculations and hypothesis testing for microbiome research. Our hope in writing this chapter is to encourage microbiome investigators to consider these statistical topics early in their design of experiments that will impact their ability to reach conclusions about their data. In particular, we focused on topics related to the data format of microbiome taxonomic sequence counts, the distribution of the data, effect size indicating how groups of microbiome samples differ from each other, and formal parametric statistical tests and power/sample size calculations. Examples for comparing diversity measures across groups using the *t*-test, the rates of occurrence of a taxon using the chi-square test, and the entire microbiome across groups using the multivariate DM distribution were presented. Short discussions were included about multiple testing adjustments (when each taxon is compared across groups separately) and the dangers in designing and analyzing data using the wrong distribution. In each example, the null hypotheses was defined, *P* values were calculated to decide if the null hypothesis should be accepted or rejected in favor of the alternative, and power/sample size calculations were performed for designing experiments.

Many investigators chose to use multiple testing (i.e., compare one taxon at a time across groups) and adjust the *P* values using one of several different multiple testing adjustment strategies.[16] This approach to data analysis has the advantage of allowing the investigator to examine individual taxa to learn how they may impact a phenotype. The shortcoming of this approach is that it ignores the interactions or correlations among the taxa and treats them as independent that is not true for metagenomic data. When the investigator is interested in discovery, this approach may be sufficient for generating hypotheses for further research. However, as the microbiome moves out of the basic science discovery phase to clinical translational research, with eventual development of biomarkers for disease and drug experiments under Food and Drug Administration guidelines, more formal statistics such as the multivariate DM test will almost surely be required.

In this chapter, because of space limitations, we have exclusively focused on parametric statistical testing and power calculations. However, other methods such as PERMANOVA[17] and ANOSIM[18] based on permutation testing are frequently used, and tools for defining effect sizes and calculating power and sample size tables are available (see, e.g., Chen *et al.*[19]). For investigators preferring to use the nonparametric approaches, there is ample literature to guide them on their use.

As biostatisticians, we are particularly interested in developing and promoting the use of parametric models for analyzing biomedical data, since these almost always have better properties than corresponding nonparametric models. For example, parametric models are almost uniformly more efficient than nonparametric models, which mean that in cases where 100 samples are required for a nonparametric model, the parametric model will have the same power with significantly fewer samples leading to cost savings in experiments. With this in mind, there are two open problems to solve with the DM test. First, if we conclude the groups are different and reject the null hypothesis, the question of which taxa are different arises. By rejecting the null hypothesis in a multivariate test, we are concluding that at least one taxon, and perhaps all taxa, are different across the groups. A *post hoc* test, analogous to those used in analysis of variance, will tell us which taxa are different. Second, many microbiome studies are longitudinal where samples are collected from the same subject at multiple times. This results in correlation of samples within subjects that must be taken into account to obtain accurate *P* values and power/sample size estimates. A generalized DM distribution model[20] can be used to solve this problem.

REFERENCES

1. Le CT. Fundamentals of biostatistical inference. CRC Press; 1992. New York, NY.
2. Hulley SB, Cummings SR, Browner WS, Grady DG, Newman TB. Designing clinical research. LWW; 2013. Philadelphia, PA.
3. Methé BA, Nelson KE, Pop M, Creasy HH, Giglio MG, Huttenhower C, et al. A framework for human microbiome research. Nature 2012;486:215–21.
4. Huttenhower C, Gevers D, Knight R, Abubucker S, Badger JH, Chinwalla AT, et al. Structure, function and diversity of the healthy human microbiome. Nature 2012;486:207–14.
5. Knight R, Jansson J, Field D, Fierer N, Desai N, Fuhrman JA, et al. Unlocking the potential of metagenomics through replicated experimental design. Nat Biotechnol 2012;30(6):513–20.
6. Cohen J. Statistical power analysis for the behavioral sciencies. Routledge; 1988. Oxford, UK.
7. Legendre P, Legendre L. Numerical ecology. Elsevier; 2012. Oxford, UK.
8. La Rosa PS, Brooks JP, Deych E, Boone EL, Edwards DJ, Wang Q, et al. Hypothesis testing and power calculations for taxonomic-based human microbiome data. PloS One 2012;7(12):e52078.
9. McMurdie PJ, Holmes S. Waste not, want not: why rarefying microbiome data is inadmissible. ArXiv e-prints; 2013. PLoS Comput Biol. Apr 2014;10(4):e1003531.
10. Steele RG, Torrie JH, Dickey D. Principles and procedures of statistics: a biometrical approach. New York: McGraw-Hill; 1980:633.
11. Shaffer JP. Multiple hypothesis testing. Annu Rev Psychol 1995;46(1):561–84.
12. Zhang J, Quan H, Ng J, Stepanavage ME. Some statistical methods for multiple endpoints in clinical trials. Controll Clin Trials 1997;18(3):204–21.
13. Holmes I, Harris K, Quince C. Dirichlet multinomial mixtures: generative models for microbial metagenomics. PLoS One 2012;7(2):e30126.
14. Chen J, Li H. Variable selection for sparse Dirichlet-multinomial regression with an application to microbiome data analysis. Ann Appl Statist 2013;7(1):418–42.
15. La Rosa PS, Deych E, Shands B, Shannon WD. HMP: hypothesis testing and power calculations for comparing metagenomic samples from HMP. R-package; 2011.
16. Young SS. Resampling-based multiple testing: examples and methods for p-value adjustment. Wiley; 1993. Oxford, UK.
17. McArdle BH, Anderson MJ. Fitting multivariate models to community data: a comment on distance-based redundancy analysis. Ecology 2001;82(1):290–7.
18. Clarke KR. Non-parametric multivariate analyses of changes in community structure. Aus J Ecol 1993;18(1):117–43.
19. Chen J, Bittinger K, Charlson ES, Hoffmann C, Lewis J,Wu GD, et al. Associating microbiome composition with environmental covariates using generalized UniFrac distances. Bioinformatics 2012;28(16):2106–13.
20. Wilson JR, Chen GS. Dirichlet-multinomial model with varying response rates over time. J Data Sci 2007;5:413–23.

CHAPTER 7

Longitudinal Microbiome Data Analysis

Georg K. Gerber

INTRODUCTION

The microbiome is inherently dynamic, driven by interactions among microbes, with the host, and with the environment. These complex dynamics begin at birth, as the infant is colonized with microbes,[1–8] and continue in the healthy adult as microbial populations vary with hormonal cycles[9] and a myriad of other host and environmental factors.[10–13] At any point in time, the microbiome can be dramatically altered, either transiently or long term, by diseases such as infections[2,14] or medical interventions such as antibiotics.[15–17] Recent advances in high-throughput experimental technologies are enabling researchers to measure dynamic behaviors of the microbiota at unprecedented scale.

Longitudinal data fundamentally provide more information than end-point data because of two special features of time-series: (a) time imposes an inherent, irreversible ordering on samples, and (b) samples exhibit statistical dependencies that are a function of time. These features of time-series data enable discovery of rich information about a system under study, including short- and long-term trends[18] and even causal interactions among system variables.[19] However, these features of time-series data also complicate analysis, necessitating the use of appropriate computational techniques.

Naïve analysis or design of time-series experiments can lead to erroneous conclusions, as illustrated in Figure 7.1. Figure 7.1A demonstrates how *aggregation over time intervals* can mask dynamic properties of data. Imagine that we are measuring diversity of a microbial ecosystem over 50 days, with the ecosystem subjected to a perturbation on day 25. If we simply average the diversity values measured at time-points prior to the perturbation, and compare with the average of the values measured after the perturbation, we will find no significant difference,

Metagenomics for Microbiology. http://dx.doi.org/10.1016/B978-0-12-410472-3.00007-5

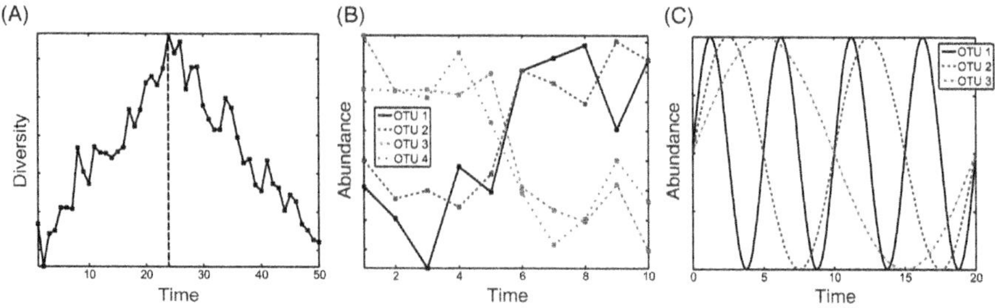

Fig. 7.1. Examples illustrating potential pitfalls when analyzing time-series data. (A) Aggregation over time-intervals can mask dynamic properties of data. In this example, ecological diversity measurements during the first half of the study have the same mean and standard deviation as those during the second half (dashed line shows mid-point of the study). (B) Aggregation of individual time-series can mask differences in temporal patterns across individuals. In this example, if the abundances of four species or OTUs are averaged at each time-point, the aggregated time-series appears flat. (C) Temporal aliasing, a form of under-sampling, can make it impossible to detect relevant changes in a system. In this example, three OTUs oscillate in abundance at different frequencies. Sampling below the frequency of oscillations (e.g., at 10-day intervals) makes it impossible to distinguish the dynamics of the OTUs.

despite the obvious opposite trends before and after the perturbation. Figure 7.1B illustrates how *aggregation of individual time-series* can mask differences in temporal patterns across individuals. Imagine that we have collected time-series data on the abundances of bacterial species or operational taxonomic units (OTUs) in the gut. Furthermore, imagine that we have data for four OTUs, in which OTUs 1 and 2 exhibit very similar patterns over time, whereas OTUs 3 and 4 exhibit mutually similar patterns but opposite to those of OTUs 1 and 2. If abundances of OTUs are summed at each time-point, the time-series would appear flat, despite the obvious changes over time for individual OTUs. Finally, Figure 7.1C illustrates how *temporal aliasing*, a form of under-sampling, can make it impossible to detect relevant changes in an ecosystem. In this example, three OTUs oscillate in abundance at different frequencies. Sampling below the frequency of oscillations (e.g., at 10-day intervals) makes it impossible to distinguish the dynamics of the OTUs.

The remainder of this chapter is organized as follows. We will first provide a set of case studies illustrating key scientific questions investigators are addressing with longitudinal microbiome analyses. Next, we will survey some computational techniques that have been used to analyze recent microbiome time-series datasets. Finally, we will conclude with a discussion of future directions in the field, including new types of longitudinal microbiome data that are becoming available, and directions for development of novel time-series analysis techniques. Note that time-series analysis is a vast subject, and its application to microbiome studies is rapidly advancing. Thus, the goal of this chapter is to provide the reader with a set of motivating examples and a conceptual background to catalyze further interest and study.

EXAMPLES

Development of the Microbiome in Early Childhood

The womb is essentially microbiologically sterile, with establishment of an individual's microbiome beginning at birth.[1–8] Infants undergo a series of developmental changes that alter their anatomy, physiology, and immune responses over time. Additionally, children's diets and environmental exposures dramatically change during the first few years of life. All these changes impact the composition of the microbiota, which, in turn, affect host metabolic and other physiological capabilities. Moreover, there is evidence from animal models that the composition of the microbiota during certain "window" periods in early life can profoundly influence immune system development.[20]

Example 1

Koenig *et al.*[2] obtained 60 fecal samples from a single infant over a 2.5-year period and used 16S ribosomal RNA (rRNA) gene sequencing to assess the microbiome. Additionally, shotgun metagenomic sequencing was performed on 12 of the samples. The phylogenetic diversity of the infant's microbiome was found to gradually increase over time, whereas the relative abundance of major taxonomic groups in the microbiota abruptly changed, particularly when solid foods were introduced. Analysis of the shotgun metagenomic data showed differences in microbial gene content between early samples, which were enriched for lactate utilization genes, and later time-points after the introduction of solid foods, which were enriched for genes associated with carbohydrate utilization, vitamin biosynthesis, and xenobiotic degradation, particularly from organisms in the Bacteroidetes phylum.

Example 2

Sharon *et al.*[5] obtained shotgun metagenomic data on 11 fecal samples collected on post-natal days 15–24 from a premature infant delivered by Cesarean section at 26 weeks gestation. Using sequence binning and genome assembly techniques that take into account temporal patterns of scaffold abundances, the investigators tracked variations over time at the levels of bacterial species and strains, and of bacteriophages. The relative abundances of three *Staphylococcus epidermidis* strains, differing in genes coding for resistance to antibiotics, heavy metals, and phage infection, were found to change over time. In addition, three bacteriophage types that infect *S. epidermidis* were identified, with the abundance of each bacteriophage type co-varying with a respective *S. epidermidis* strain. A novel *Propionibacterium* species was also identified, which was present late in the time-series and contained genes coding for inositol and sialic acid metabolism not present in *Propionibacterium* species seen early in the time course.

Microbiome Variability Over Time in Healthy Adults

The microbiome does not become static after childhood. Healthy adults routinely engage in behaviors that can alter their microbiomes, including eating different foods day to day[12] and coming into contact with new reservoirs of commensal microbes through travel.[13] Furthermore, both women and men experience hormonal cycling and other time-varying physiology that influence their microbiomes.[9] Characterization of variations in the microbes of healthy adults over time can provide insights into the factors that drive temporal microbiome variability, as well as lay the foundation for discrimination of normal temporal microbiome variability from dysbiosis.

Example 3

Caporaso *et al.*[21] obtained fecal, oral, and skin samples from two healthy adults with a mean interval of 1.12 days between samples, for 6 months in one subject and 15 months in the other. A total of 396 time-points were sampled and microbiomes were assessed using 16S rRNA gene sequencing. Individuals' microbes, in terms of presence/absence and abundance of OTUs, were found to vary considerably across months, weeks, and even days. Many taxa persisted for multiple time-points, but few were present over the entire time-course. However, microbiomes in different body sites or across individuals remained distinguishable, suggesting that microbial communities throughout the body change over time but maintain body site-specific and host-specific distinctions.

Example 4

Gajer *et al.*[9] obtained vaginal swabs twice weekly over a 16-week period from 32 healthy reproductive-age women and assessed their microbiomes using 16S rRNA gene sequencing. Five types of vaginal bacterial communities were identified, with most communities dominated by a particular *Lactobacillus* species. Changes in community type over time were found to be complex and individualized, but exhibited certain commonalities. For instance, communities dominated by *L. gasseri* rarely transitioned to other community types, whereas communities dominated by *L. crispatus* often transitioned to communities dominated by *L. iners*. Increased variability in the vaginal microbiome was found to be associated with specific times in the menstrual cycle, bacterial community composition, and sexual activity. However, women in the study remained healthy despite changes in the variability of their vaginal microbiomes, suggesting that there is a substantial range of "normal" microbiome variability within and across the human population.

Responses of the Microbiota to Perturbations

Infections[2] and other illnesses, or intentional interventions such as antibiotic therapies[15] or dietary modifications[22] can dramatically alter the microbiota. A key question is whether the microbiome recovers to its original state after the perturbation or ends up in a new state, and how quickly equilibration occurs. Knowledge of the effects of perturbations on the microbiome can help us to understand the robustness of healthy or dysbiotic microbiota to changes induced by environmental interactions or medical interventions, and ultimately provide insights into how we can reshape the microbiome to benefit the host.

Example 5

Dethlefsen and Relman[15] obtained 52–56 stool samples from three healthy human subjects over a 10-month period during which each subject received 2 5-day courses of oral antibiotics. Composition of the microbiota was assessed using 16S rRNA gene sequencing. Microbial diversity was seen to decline rapidly in all subjects within 3–4 days of initiating antibiotics, with microbial diversity generally recovering within a few days of ceasing antibiotics. Some commonalities were seen in post-antibiotic alterations to the microbiota of all subjects, such as decreases in abundances of OTUs in the family *Ruminococcaceae* and

Lachnospiraceae. However, alterations substantially varied across subjects and between each antibiotic course in a single subject. Furthermore, although the composition of the gut microbiota was qualitatively observed to stabilize by the end of the experiment, it remained altered relative to its initial composition.

Example 6

Wu *et al.*[23] performed a controlled diet experiment, in which 10 human subjects were sequestered for 10 days in a hospital environment and randomized to receive either a high-fat/low-fiber or a low-fat/high-fiber diet. Stool samples were collected daily from the subjects, and the composition of the microbiota was assessed using 16S rRNA gene sequencing. Changes in overall microbial community structure were detected within 24 hours of initiating the controlled diet, although changes in abundances of taxa tended to be individualized and few commonalities were detected at this level of analysis across subjects.

COMPUTATIONAL METHODS FOR ANALYZING MICROBIOME TIME-SERIES DATA

Regression-Based Techniques

Long-term dependencies over time, or trends, can be modeled by regressing a series of observations on time.[18] That is, we can model a series of observations (dependent variables), such as relative abundances of an OTU over time or ecological diversity of the gut microbiota over time, as a function of time and other covariates (independent variables). Such models have been successfully applied to analyzing microbiome data, such as the study by Gajer *et al.*[9] (Example 4), which used a regression model to evaluate the dependence of the human vaginal microbiome on time in the menstrual cycle and other covariates.

The regression of a series of observations Y on time can be expressed with the following general equation:

$$Y_t = f(t;\theta) + \varepsilon_t \tag{1}$$

Here, Y_t represents the value of the dependent variable at time-point t, $f(t;\theta)$ is a function of time with parameter vector θ, and ε_t is a random error term (e.g., normally distributed noise). By specifying different functional forms for $f(t;\theta)$, the general model can capture many types of trends in time-series data. Practically, the input to such a regression model is a column of observed values (e.g., OTU abundances) and a column of the times at which the values were observed. A statistics software package, such as R (Free Software) or Matlab (commercial software from MathWorks, Natick, MA) can then be used to fit the model.

Figure 7.2A and B depicts a simple linear and a periodic or cyclical trend, respectively. If the functional form for $f(t;\theta)$ is unknown *a priori*, various approaches can be used. For instance, as shown in Figure 7.2B, $f(t;\theta)$ can be expressed as a flexible spline,[24,25] which consists of a series of low-degree polynomial segments defined piecewise but smoothly joined over the time-series; splines will fit data better than a single high-degree polynomial function, which may dramatically under- or overshoot the data.

An alternative type of regression model, termed the autoregressive (AR) model, does not directly regress observations on time as in Equation 1, but instead regresses present observations on prior observations. Conceptually, AR models allow prediction of the future and capture the phenomenon of increasing uncertainty about events further in the future; in contrast, models that directly regress on time assume an equal ability to predict a variable at any time-point. For these reasons, AR models are extensively used to analyze complex, noisy data, particularly economic measurements such as stock prices.[18] AR techniques have also

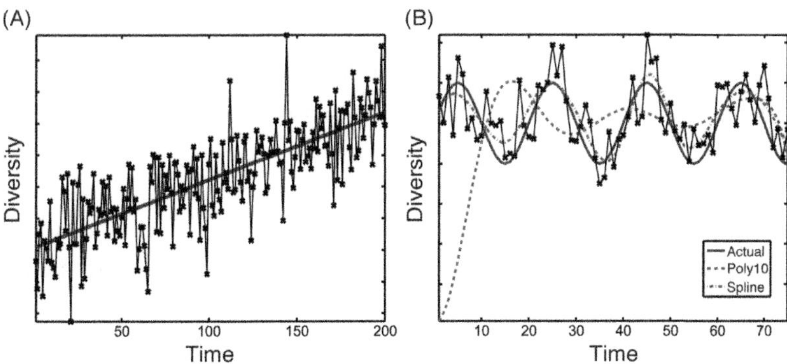

Fig. 7.2. Regression models with time as the independent variable can capture trends in longitudinal data. (A) Example of a simple trend in ecological diversity over time (x's joined with a thin line) fit with a linear regression model (thick solid line). (B) Example of a cyclical trend of ecological diversity over time (x's joined with a thin line depict data, thick solid line depicts actual trend) fit with a polynomial of degree 10 ("poly10," dashed line) or a spline (dot-dashed line).

been used in several microbiota studies, such as Palmer *et al.*,[3] who used these techniques to assess the tendency of different taxonomic groups of bacteria to persist once established during colonization of the gut during childhood.

A general formula for an AR model is:

$$Y_t = \sum_{i=1}^{p} \theta_i Y_{t-i} + \varepsilon_t \tag{2}$$

This equation specifies an autoregressive model of *order p*, denoted $AR(p)$, meaning that the data at time t depends on p prior data points. As in Equation 1, Y_t represents the value of the variable of interest at time-point t, for example, the relative abundance of a particular OTU. The right-hand side of Equation 2, however, differs from that of Equation 1, in that time is not explicitly represented. Rather, time is captured by the past values of the variable of interest.

Figure 7.3A and C depicts data simulated from example AR models. In an $AR(1)$ model (Figure 7.3A), the signal at adjacent time-points tends to be quite similar, but these similarities rapidly dissipate over time. An $AR(20)$ model (Figure 7.3C), in contrast, exhibits long-term dependencies or trends over time. Autocorrelation analyses (Figure 7.3B and D), which involve calculating the correlation between a time-series (e.g., series of relative abundances of an OTU) and a lagged version of itself at various lags, provide a quick means to explore the order of AR model needed to capture the dependencies present in a time-series. Statistics software packages, such as R and Matlab, provide various functions for readily fitting AR models and performing autocorrelation analyses and visualizations.

State-Space Models

Probabilistic state-space models assume that the outputs or measurements of a system depend on its state, which can change over time. In some cases, a system's state may directly correspond to observable quantities such as its temperature. However, in many cases, a system's state cannot be directly observed, and must be inferred from its outputs. State-space models are particularly useful for detecting when a system undergoes a substantial shift. Additionally, these models are useful for

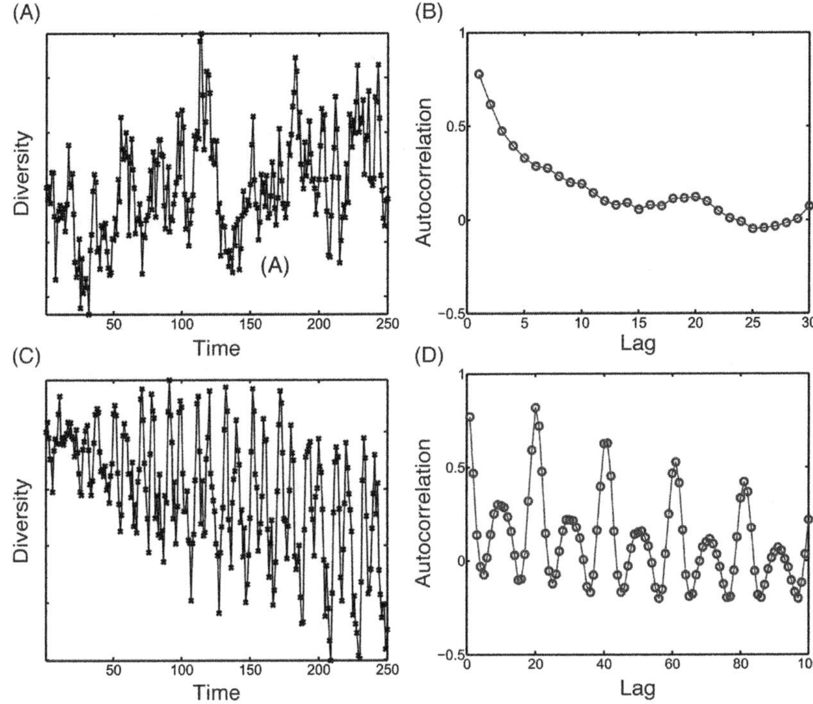

Fig. 7.3. Autoregressive (AR) models can capture dependencies in longitudinal data at varying time-scales. (A) Example of an AR(1) *(order 1) model for ecological diversity. (B) Autocorrelation plot for the* AR(1) *model, showing exponential decay of correlations between measurements over time. (C) Example of an* AR(20) *(order 20) model for ecological diversity. (D) Autocorrelation plot for the* AR(20) *model, showing long-term periodic correlations between measurements.*

analyzing multiple variables over time, because the underlying state of a system can simultaneously drive the behavior of many variables.

Gajer *et al.*[9] (Example 4) informally used a state-space model-like approach to analyze human vaginal microbiome time-series data. Different representative microbial community compositions were identified in a preprocessing step, and these compositions were defined as the underlying system states. It was then determined which of the predefined states was most similar to a subject's microbiota at each time-point sampled, and the frequency of transitions between states was analyzed.

A widely used formal state-space model is the hidden Markov model (HMM),[26] which assumes that a system is in a discrete state at any given time and the system may probabilistically change state at each discrete time-step. States are "hidden" in the sense that they are not directly observed but can be inferred from data using efficient algorithms. The order p of an HMM denotes the number of past states that the present state depends on. The number of states must be prespecified for standard HMM models, although nonparametric Bayesian HMMs have been developed that can infer the number of states from the data.[27]

Figure 7.4 provides an example of a five-state HMM of order 1. States are depicted as circles. The numbers next to arrows indicate the probability of transitioning between states at each time-step. For instance, there is a 3% probability of transitioning from state 1 to state 2 at each time-step. Figure 7.4B illustrates a trajectory, or series of states,

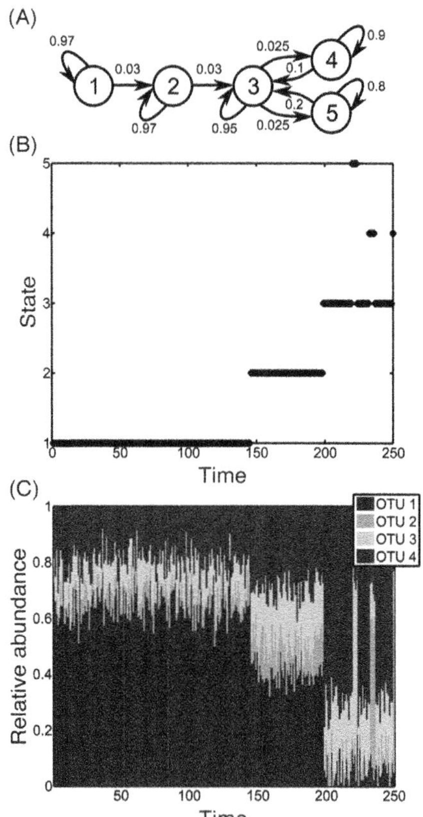

Fig. 7.4. Hidden Markov Models (HMMs) can capture shifts over time in the underlying state of a system. (A) Example of an order one HMM with five states. Circles denote states and labeled arrows indicate transition probabilities. The observation model is not shown. (B) A trajectory of states simulated from the HMM. (C) Data simulated for four OTUs from the trajectory of states in panel (B).

simulated from the example HMM. These states would not be directly observed. Instead, as shown in Figure 7.4C, we might have data consisting of a time-series of relative abundances of OTUs. The microbial community composition corresponding to each state, the transition probabilities between states, and the state at each time-step could then be inferred from the available time-series data.

Temporal Pattern Clustering

Groups of microbial species have been observed to exhibit similar patterns of changes in relative abundance over time within complex host–microbial ecosystems.[28] These groups or clusters may be comprised of organisms with similar metabolic or other functional capabilities, or organisms that are physically proximate in microenvironments within the host. The number of such clusters within the microbiota of different individuals or within the same individual subjected to different perturbations can inform us as to the repertoire of responses available within a microbiome. Computational methods that cluster data often require the user to prespecify the number of clusters. However, in many cases, including for most microbiome applications, the number of clusters present in data is unknown *a priori*. Thus, automated and statistically principled clustering methods are crucial for these applications.

For example, we developed the Microbiome Counts Trajectories Infinite Mixture Model Engine (MC-TIMME),[28] a time-series clustering algorithm specifically tailored for analyzing microbiome data that automatically infers the number of temporal patterns from the data. MC-TIMME uses a nonparametric Bayesian technique, the Dirichlet Process,[29] which assumes that the data arise from an unlimited (infinite) mixture of continuous-time temporal patterns. Using approximate inference methods, the fully Bayesian MC-TIMME algorithm estimates the distribution over model variables, including the number of nonempty mixture components. MC-TIMME thus provides "error bars" (measures of uncertainty) over all variables, including the number of clusters/temporal patterns and the shape of each temporal pattern. Additionally, MC-TIMME provides more accurate estimates of individual temporal patterns, by aggregating information across multiple time-series exhibiting similar patterns.

In a proof-of-principle application, we re-analyzed the antibiotic perturbation data from Dethlefsen and Relman[15] (Example 6) using

MC-TIMME and established several new results. We introduced a new measure, called signature diversity, that quantifies the number of different temporal patterns in a perturbed host–microbial ecosystem and demonstrates marked similarities in signature diversity across subjects despite differences in standard measures of ecological diversity or microbial community composition. In addition, our continuous-time model of temporal patterns allowed us to accurately quantitate the time for individual OTU's relative abundances to equilibrate after antibiotic exposure and showed that OTUs generally equilibrated more quickly after the second antibiotic exposure. MC-TIMME also allowed us to infer Consensus Signature Groups (CSGs) or sets of OTUs exhibiting similar temporal patterns. By ordering the CSGs by how quickly their component OTUs equilibrated after the first antibiotic exposure, we found a timeline of successive changes subsequent to antibiotic exposure in subcommunities of microorganisms in the human gut, with CSGs having different taxonomic compositions with distinct physiologic capabilities such as acetate or butyrate production.

Automated Experimental Design

Principled experimental design is particularly important for longitudinal studies. As illustrated in Figure 7.1C, under-sampling in a longitudinal study, on one hand, can make important changes in the system undetectable. On the other hand, oversampling can be prohibitively expensive or logistically impractical, particularly when human subjects are involved. Often, the frequency at which changes may occur in a complex host–microbial ecosystem over time is unknown. In these cases, pilot experiments, with frequent sampling, will need to be performed to collect preliminary data on which to base an experimental design to study a larger cohort.

We developed an automated experimental design technique for longitudinal microbiome studies in tandem with the MC-TIMME[28] algorithm described above. Our technique formulates the experimental design task as an information theoretic optimization problem,[30] in which the objective function to be optimized is the expected information gain (or reduction in entropy) achieved over all possible observations at a set of time-points. Data from a pilot experiment is used to estimate the distribution over possible observations in future experiments. In general, our method will favor experimental designs that sample at time-points with the greatest uncertainty or when the system under study is predicted to change most rapidly. In a proof-of-principle demonstration, we applied our method to data collected by Dethlefsen and Relman[15] (Example 6), measuring changes in the microbiota of human subjects given pulses of oral antibiotics. Our method generated an optimal experimental design that agreed with the original design at many time-points. However, the optimal design had some notable differences, such as denser sampling for approximately 2 weeks after the first antibiotic exposure, which was consistent with our analyses demonstrating that some groups of OTUs took considerably longer than others to reach equilibrium abundances after the first antibiotic exposure.

CONCLUSION

In this chapter, we introduced the topic of longitudinal microbiome data analysis through examples of studies from the literature and a brief overview of some important computational techniques. The microbiome field is rapidly advancing, as new technologies become available and investigators creatively apply these technologies to address basic scientific or clinically relevant questions. Example studies covered in this chapter focused on analyzing abundances of microbes or the gene content of microbial communities over time, using marker gene or shotgun metagenomics sequencing strategies. Studies measuring functional properties of the microbiota over time,[16] such as transcriptomes, metabolomes, and proteomes, hold great promise. Another important emerging focus in the field are studies that prospectively analyze the microbiota of human subjects at risk for developing a disease.[6,22,31] The field is also moving beyond assaying only bacterial populations to examining populations of other microorganisms including viruses,[32] which exhibit astounding genetic plasticity over time.

To date, few computational tools have been developed that are specially tailored for analyzing microbiome time-series data. Special properties of these data include their discrete nature (counts), high levels of noise, large numbers of variables with dependencies due to phylogenetic and functional relationships, and complex dynamics due to multiple types of interactions including microbe–microbe, host–microbe, and environment–microbe relationships. There is a tremendous practical need for new computational techniques that appropriately model these properties, and moreover, development of such techniques will undoubtedly provide the impetus for new, broadly applicable theoretical advances in computer science, statistics, and mathematics.

REFERENCES

1. de Muinck EJ, Lagesen K, Afset JE, et al. Comparisons of infant Escherichia coli isolates link genomic profiles with adaptation to the ecological niche. BMC Genomics 2013;14(1):81.
2. Koenig JE, Spor A, Scalfone N, et al. Succession of microbial consortia in the developing infant gut microbiome. Proc Natl Acad Sci USA 2011;108(Suppl):4578–85.
3. Palmer C, Bik EM, DiGiulio DB, Relman Da, Brown PO. Development of the human infant intestinal microbiota. PLoS Biol 2007;5:e177.
4. Schloss PD, Schubert AM, Zackular JP, Iverson KD, Young VB, Petrosino JF. Stabilization of the murine gut microbiome following weaning. Gut Microbes 2012;3(4):383–93.
5. Sharon I, Morowitz MJ, Thomas BC, Costello EK, Relman DA, Banfield JF. Time series community genomics analysis reveals rapid shifts in bacterial species, strains, and phage during infant gut colonization. Genome Res 2013;23(1):111–20.
6. Stewart CJ, Marrs EC, Nelson A, et al. Development of the preterm gut microbiome in twins at risk of necrotising enterocolitis and sepsis. PLoS One 2013;8(8):e73465.
7. Trosvik P, Stenseth NC, Rudi K. Convergent temporal dynamics of the human infant gut microbiota. ISME J 2010;4(2):151–8.
8. White RA, Bjornholt JV, Baird DD, et al. Novel developmental analyses identify longitudinal patterns of early gut microbiota that affect infant growth. PLoS Comput Biol 2013;9(5):e1003042.
9. Gajer P, Brotman RM, Bai G, et al. Temporal dynamics of the human vaginal microbiota. Sci Transl Med 2012;4(132):132ra52.
10. Claesson MJ, Cusack S, O'Sullivan O, et al. Composition, variability, and temporal stability of the intestinal microbiota of the elderly. Proc Natl Acad Sci USA 2011;108(Suppl 1):4586–91.
11. Turnbaugh PJ, Hamady M, Yatsunenko T, et al. A core gut microbiome in obese and lean twins. Nature 2009;457:480–4.
12. Turnbaugh PJ, Ridaura VK, Faith JJ, Rey FE, Knight R, Gordon JI. The effect of diet on the human gut microbiome: a metagenomic analysis in humanized gnotobiotic mice. Sci Transl Med 2009;1(6). 6ra14.
13. Yatsunenko T, Rey FE, Manary MJ, et al. Human gut microbiome viewed across age and geography. Nature 2012;486:222–7.
14. Hoffmann C, Hill DA, Minkah N, et al. Community-wide response of the gut microbiota to enteropathogenic Citrobacter rodentium infection revealed by deep sequencing. Infect Immun 2009;77(10):4668–78.
15. Dethlefsen L, Relman DA. Incomplete recovery and individualized responses of the human distal gut microbiota to repeated antibiotic perturbation. Proc Natl Acad Sci USA 2011;108(Suppl 1):4554–61.
16. Perez-Cobas AE, Gosalbes MJ, Friedrichs A, et al. Gut microbiota disturbance during antibiotic therapy: a multi-omic approach. Gut 2012;62(11):1591–601.

17. Peterfreund GL, Vandivier LE, Sinha R, et al. Succession in the gut microbiome following antibiotic and antibody therapies for Clostridium difficile. PLoS One 2012;7(10):e46966.

18. Wei WWS. Time series analysis: univariate and multivariate methods. 2nd ed. New Jersey: Pearson; 2005.

19. Sugihara G, May R, Ye H, et al. Detecting causality in complex ecosystems. Science 2012;338(6106):496–500.

20. Hooper LV, Littman DR, Macpherson AJ. Interactions between the microbiota and the immune system. Science 2012;336:1268–73.

21. Caporaso JG, Lauber CL, Costello EK, et al. Moving pictures of the human microbiome. Genome Biol 2011;12(5):R50.

22. Smith MI, Yatsunenko T, Manary MJ, et al. Gut microbiomes of Malawian twin pairs discordant for kwashiorkor. Science 2013;339(6119):548–54.

23. Wu GD, Chen J, Hoffmann C, et al. Linking long-term dietary patterns with gut microbial enterotypes. Science 2011;334(6052):105–8.

24. Bar-Joseph Z, Gerber G, Simon I, Gifford DK, Jaakkola TS. Comparing the continuous representation of time-series expression profiles to identify differentially expressed genes. Proc Natl Acad Sci USA 2003;100(18):10146–51.

25. Eilers PHC, Marx BD. Flexible smoothing with B-splines and penalties. Statist Sci 1996;11:89–121.

26. Rabiner LR. A tutorial on hidden Markov models and selected applications in speech recognition. Proc IEEE 1989;77:257–86.

27. Fox EB, Sudderth EB, Jordan MI, Willsky AS. A sticky HDP-HMM with application to speaker diarization. Ann Appl Statist 2011;5(2A):1020–56.

28. Gerber GK, Onderdonk AB, Bry L. Inferring dynamic signatures of microbes in complex host ecosystems. PLoS Comput Biol 2012;8(8):e1002624.

29. Rasmussen C. The Infinite Gaussian Mixture Model. Advances in neural information processing systems (NIPS). Cambridge, MA: MIT Press; 2000. pp. 554–560.

30. Lindley DV. On a measure of the information provided by an experiment. Ann Math Statist 1956;27(4):986–1005.

31. Jenq RR, Ubeda C, Taur Y, et al. Regulation of intestinal inflammation by microbiota following allogeneic bone marrow transplantation. J Exp Med 2012;209(5):903–11.

32. Minot S, Bryson A, Chehoud C, Wu GD, Lewis JD, Bushman FD. Rapid evolution of the human gut virome. Proc Natl Acad Sci USA 2013;110(30):12450–5.

CHAPTER 8

Metagenomics for Bacteriology

Erika del Castillo and Jacques Izard

The study of bacteria, or bacteriology, has gone through transformative waves since its inception in the 1600s. It all started by the visualization of bacteria using light microscopy by Antonie van Leeuwenhoek, when he first described "animalcules." Direct cellular observation then evolved into utilizing different wavelengths on novel platforms such as electron, fluorescence, and even near-infrared microscopy. Understanding the link between microbes and disease (pathogenicity) began with the ability to isolate and cultivate organisms through aseptic methodologies starting in the 1700s. These techniques became more prevalent in the following centuries with the work of famous scientists such as Louis Pasteur and Robert Koch, and many others since then. The relationship between bacteria and the host's immune system was first inferred in the 1800s, and to date is continuing to unveil its mysteries. During the last century, researchers initiated the era of molecular genetics. The discovery of the first-generation sequencing technology, the Sanger method, and, later, the polymerase chain reaction technology propelled the molecular genetics field by exponentially expanding the knowledge of relationship between gene structure and function. The rise of commercially available next-generation sequencing methodologies, in the beginning of this century, is drastically allowing larger amount of information to be acquired, in a manner open to the democratization of the approach.

HEALTHY HOSTS AND MICROBIOMES

Cooperation and association, in their broadest meanings, are ubiquitous and part of the evolutionary processes between bacteria and host. This mutually beneficial association has sustained coevolution through different habitats.

Microbiota–host cooperation starts from the moment development begins in the environment outside of the genetic progenitors, for example, the

Metagenomics for Microbiology. http://dx.doi.org/10.1016/B978-0-12-410472-3.00008-7

microbiota changes from postlarvae stage to the adult stage in an oyster, throughout the different stages of metamorphosis for the frog, and from birth to adulthood for mammals.[1–4] Interestingly, it seems that individual-specific strains, when established, are stable in an environment even if their relative abundance changes over time.[5,6]

To redefine the concept of health, the Human Microbiome Project (HMP) consortium recruited subjects without sign of proinflammatory condition or disease.[7,8] The studies from the acquired metagenomic data sets, from multiple body sites, show that diversity of microbes is key to health.[7,8] Other studies have shown that the microbiome influences a wide spectrum of biological events including the immune function and behavior of the host.[9–12]

Our life expectancy has drastically improved in the last 100 years. The impact of these changes on the ancestral mutualistic relationships between humans and microbes has to be part of those progresses but is not well understood. A study on calcified dental plaque has shown that from the Neolithic (remains dated 7550–5450 years before present) to the medieval times, the oral microbiota was more diverse than the present oral microbiota and was relatively stable.[13] A study of 1400-year-old coprolites from northern Mexico shows a more diverse gut microbiota compared with those of modern urbanized populations, however, more similar to rural populations with different modern life-styles.[14] Many questions remain as we are just at the beginning of our understanding on how our own microbiomes are key to our survival.

WHAT ABOUT DISEASE?

Human diseases are not a new burden. At a middle-age monastic site in Germany, adult skeletons were recovered with evidence of mild-to-severe periodontitis (oral microbial infection leading to tooth loss). Using DNA extracted from the teeth of the skeletons, researchers were able to reconstruct the genome of a known pathogen, *Tannerella forsythia*,[15] and identify the molecular signatures of other periodontitis-associated species.

The treatment of disease has been an interest of any society, and microbial modification has always been a component of treatment. While plant-based therapy was probably the way to treat diseases in Neolithic times, refined chemical compounds are now available as pharmaceuticals.

Regardless of the source, the microbiome can be targeted by these antimicrobials modifying community structure and metabolic potential.[16,17] Next-generation sequencing is providing a greater depth of understanding of the broader effect during treatment as well as host microbiome recovery post-treatment.[18]

Medical challenges where antimicrobial therapy has been unsuccessful have led to new approaches, such as fecal transplants. Refractory recurrent *Clostridium difficile* infections do not respond to appropriate antibiotic therapy. Fecal transplants offer the possibility of a rapid remodeling of the receiver gut microbiome toward its donor's transplant profile, and at the same time eliminate *C. difficile* challenge.[19,20]

Treatment successes and failures might have to be revisited in the context of the host–microbiome relationship. Therapeutic drugs alter the host–microbiota composition and can colocalize specific bacteria to lympoid tissue or cells where they can synergistically modulate and influence the efficacy of the therapeutic drugs.[21] Thus, in addition of being the target, the microbiome can also act as a modulator of treatment efficacy by altering the expected effect.[21,22] A thorough understanding of the molecular bases of host–microbiota interactions could lead to the development of new therapeutic strategies for treating human disorders, as well as decreasing the toxicity of some of the present treatments.

While new approaches are being designed, the realms of traditional eastern and western medicine are slowly beginning to intersect with our increased understanding of the microbiome role in health and disease. Traditional Chinese medicine has been widely used for millennia in the treatment of various diseases in East Asian countries. The analysis of tongue coating, a fundamental practice in Chinese medicine, has been used as a basis to differentiate the microbiota in the case of hot and cold syndromes.[23] The observed differences suggest that tongue-associated microbiomes could be used as a novel holistic biomarker to subtype human host populations.

FOOD, BIOTRANSFORMATION, AND LIFE

Since food and nutrition are essential to the survival of all living beings on Earth, it comes as no surprise that the first metagenomic studies have focused on the gut microbiota. As the body of publication is significant, we will look at two cross-pollinations among fields.

The comparative genomic analysis of the genome of the giant panda uncovered the presence of the enzymes associated with a carnivorous digestive system while lacking the enzymes to digest cellulose, the principal component of their bamboo diet. The apparent metabolic contradiction was resolved while studying their gut microbiome. The study shows that *Clostridia* bacteria appear to be the microbial symbionts bridging this necessary metabolic gap.[24] Without the presence of *Clostridia* in the gut microbiome, the panda would not be able to survive on a diet of bamboo. The presence of stable and specific cellulose-degrading species in gut microbiome has allowed the giant panda to transition from a carnivore to a herbivore life-style, illustrating a coevolutionary process between the host and its gut microbiome.

This importance in energy balance has been underlined in metabolic transfer from bacteria to the host in obesity, in voluntary diet modification, as well as in the forced change of diet due to habitat loss.[25–27] In both humans and mouse models, it has been shown that changes to the gut ratio of Bacteroidetes/Firmicutes modulate the capacity for energy harvest, with a decrease of Bacteroidetes being associated with obesity. This correlation allowed for a better understanding of the physiology of the Australian sea lion metabolism. Their gut has a dominant composition of Firmicutes predisposing this aquatic mammal toward an excess of body fat needed for thermoregulation within their cold oceanic habitat.[25,28,29] Microbiota balance or dysbiosis depends on the context and physiology of the host. The numbers of bacteria or genes by themselves do not provide a complete story: a larger-scale analysis is required to understand the intricacy of the microbiome relationships sustaining life.[25–27]

SOME PRACTICAL USAGE OF THE MICROBIOMES

The utilization of bacteria in food production by many societies/civilizations/cultures predates modern microbiology. In Asia, before the end of the first millennia AD, a low-temperature lactic acid-based fermentation process was used to preserve food for the winter season. Now kimchi is known worldwide. Metagenomic analysis of the kimchi fermentation process led to a greater understanding of microbial community composition, pH, and respiration-associated function modulation during this month-long process.[30]

During the middle ages, the Europeans developed the process to produce the cheese products that we still enjoy. Today, the Italian Mozarella, Grana Padano, and Parmigiano Reggiano cheeses, while from different geographic regions, are all produced by microbial communities with similar metabolic functionality, composed of thermophilic, aciduric, and moderately heat-resistant lactic acid bacteria.[31] A few additional examples of the ubiquitous use of microbiota in food are the preparation of cocoa bean in the Americas, the fermentation of millet to make boza drink in the Middle East, and the fermentation of teff to make the sourdough-risen flatbread injera in Africa.

The soil microbiome around the plant rhizosphere is modified by plant roots exudates. In agriculture, metagenomics approaches offer the potential to modify soil microbiome structure using blends of phytochemicals that might support beneficial microbiota with the goal of enhancing crop yields, sustainability, and fend off infections by maximizing a healthy plant–soil interaction.[32]

In aquaculture, metagenomics approaches can help in the design of preventive strategies with the goal of enhancing the health of the fishes by the manipulation of their gut microbiota. Recently, the gut microbiota of commercially valuable warm-water fishes, including the channel catfish and the largemouth bass, has been characterized with the goal of growth optimization and disease control.[33]

OUR SOCIETAL CHOICES INFLUENCE THE MICROBIOMES

As we move toward a better understanding of the intersection of human behaviors (both individual and societal), the human microbiome, and the environments in which humans live, the overall complexity drastically increases. The choices we make either as an individual or as a society influence our interactions with the diverse microbiomes surrounding us. Furthermore, the impact is not limited to us and can be positive, neutral, or negative to others. For example, the microbial communities in the drinking water distribution systems depend on the source of water, the tubing material (copper, stainless steel, or polyvinyl chloride), and the regimen and selection of disinfection methods on drinking water by private and municipal water services leading to a safe drinking water.[34–36] Although the microbiota present in the drinking water sources might be regionally or locally determined, the need for clean and safe drinking water is universal.

Architectural choices of our homes, schools, and hospitals, by the design of the airflow, the temperature, the relative humidity, and the interactive surfaces in the different sections of the rooms or buildings, influence the surrounding surface and aerosol microbiota.[37,38] Our choice of mode of transportation, whether private or public, also has an influence that might be as equivalent to our exposure to the outdoor conditions from the same environment, showing that safety exist also in numbers.[39,40]

At a larger scale, how different societies use the land and water resources can have long distance and long-term effects in the microbiome of those environments. Hurricanes, for example, are able to aerosolize a large amount of microbial cells to the upper troposphere that can potentially influence the hydrological cycle, clouds, and climate.[41] The microbiome–society interaction is bidirectional and until recently we have been largely blind to this relationship.

Recent developments create greater optimism for a better management of our inner ecology as well as the biosphere. These events include a wider spread of scientific theories, as shown by the large number of individuals taking online scientific courses,[42,43] the increasing strength of citizen science,[44,45] and a greater access to scientific tools through open-source software and scientific literature from open-access publishing.[46]

DIVING INTO A DETAILED VIEW OF THE SCALES INVOLVED

Looking at a smaller scale, the coexistence of microorganisms in communities, microbial networking, and community development are at the center of the dynamic aspect of the microbiomes. Bioinformatic approaches are allowing us to redefine our understanding of the relationships between members within the communities, as well as the rules of association, competition, and exclusion.

Metagenomic approaches are finally allowing an in-depth comparative analysis of multiple sites within an individual and across populations. The first large-scale effort of this type was performed in the

Sargasso Sea at different oceanic sampling stations.[47] More recently, in the cohort of the HMP, a study of 18 body sites was performed, and later was complemented by an additional selection reaching an overview of 22 human body habitats.[7,48] This biogeography is associated with the presence of relationship networks of diverse structures. Traditional microbiology has shown that these relationships can lead to direct physical interactions associated with the succession of biofilm formation, ultimately leading to an interactome.[49] When analyzing next-generation sequencing data, this network expands to co-occurrence networks, where phylotypes are typically, but not always, present together at a site.[8,23,50] Although we are far from understanding all of these relationships, a metabolic interdependence exists, because of a degradation cascade of nutriments that affect both the microbiome and the host.

Within a microenvironment, horizontal gene transfer seems to be a competitive option to complement the panel of functional capabilities, as shown by the analysis of available genomes.[51] In the specific case of the human gut bacterium *Bacteroides plebius*, the genetic exchange occurred with a marine bacteria. This gene transfer facilitates seaweed digestion in some Japanese individuals carrying *B. plebius* enhanced by this genetic addition.[52] Another available option in multispecies communities is to use mutualistic cooperation to both enhance nutriment intake and protecting themselves from the host.[41,42,53,54]

The complexity of the interactions becomes more apparent as we go deeper into the details of the massive data sets. The initial findings on the gut microbiome, from the MetaHIT project, indicates that microbial genes outnumber human genes by more than 100-fold, predicting over 3 million bacterial genes in the gut alone.[55] Multiple scales of observation are needed, from the atomic structure modification of proteins during an enzymatic digest to the gradient of molecules within the cell, the chemotactic abilities of cells to improve their nutrient uptake or flee toxics, the surface protein providing direct interaction with other cells and to the assemblage of cells forming biofilms, and the surface to which the biofilm associates. These integrated scales of interactions, mechanistic events, and optimizations are crucial for survival, dormancy, or ability to thrive. It is up to us to understand the rules that have been in place for million of years.

WHAT WOULD HELP TO FURTHER THE LEAP?

Metagenomics heavily relies on reference databases to improve the analysis phase for phylogenetic, metabolic, and functional content including hypothetical small RNAs and proteins. Assessing the biodiversity in greater details also presents the challenge of validation in the laboratory as it is a more controlled environment.

Bacterial Systematics

For over 140 years, the world of bacterial systematics has been evolving because of technological and conceptual advances.[56] As of 2013, the number of validly named taxa rose to about 2000 genera and 10,600 species from 29 phyla (list graciously maintained by Dr Euzéby, available at www.bacterio.net). To this list, additional organisms deposited in culture collections are awaiting naming after isolation and genome sequencing during large-scale efforts such as the HMP[57,58] (list available at www.hmpdacc.org). Beyond traditional methods, whole-genome study allows proper positioning in the phylogenetic hierarchy. However, the move to whole-genomes phylogenetic analysis has been curbed, until recently, by the limited number of whole-genome and high-quality genomic sequence drafts. Additionally, new tools need to be developed to go further and define strain-level phylogeny based on genetic content.[59] This will undoubtedly bring some conflicts with the present classification as it happened when the 16S rRNA gene phylogenetic classification competed with the phenotypic classification.[56] Concurrently, databases such as the Ribosomal Database Project, Greengenes, SILVA, Human Oral Microbiome Database, and others expand beyond officially named bacteria and maintain our ability to do 16S rRNA gene phylogenetic analyses.[60–63]

Bacterial Quantitation

Refined quantitative analysis to study the relative abundance of different bacteria will have to take into account the copy number of genes including the 16S rRNA gene. As shown in Table 8.1, the number of 16S rRNA genes can vary from 1 to 15 with no specific correlation to genome size, GC%, or membership to a specific genus or phylum. For example, two strains of the Firmicutes *Bacillus subtilis* differ by two copies (8 vs. 10), and their genome size by 4% (Table 8.1). Within the Proteobacteria,

Table 8.1 16S rRNA Gene Copy Numbers in a Subset of Bacterial Genomes

Phylum	Average 16S rRNA Gene Copies in Phylum[a]	Organism Name	16S rRNA Gene Copy Number[b]	Genome Size (bp)[d,e]	GC%[e]
Actinobacteria	3.1 ± 1.7	*Frankia sp.* Cc13	2	5,433,628	70.1
		Frankia sp. EuI1c	3	8,815,781	72.3
		Kineococcus radiotolerans SRS30216	4	4,956,672	74.2
Bacteroidetes	3.5 ± 1.5	Candidatus *Sulcia muelleri* DMIN	1	243,933	22.5
		Tannerella forsythia ATCC 43037	2	3,405,521	47.0
		Porphyromonas gingivalis ATCC 33277	4	2,354,886	48.4
Cyanobacteria	2.3 ± 1.2	*Synechocystis sp.* PCC 6803	2	3,947,019	47.3
Deinococcus-Thermus	2.7 ± 1.0	*Thermus thermophilus* HB-8	2	2,116,056	69.5
		Deinococcus radiodurans R1	3	3,284,156	66.6
Firmicutes	5.8 ± 2.8	*Lactobacillus casei* ATCC 334	5	2,924,325	46.6
		Staphylococcus aureus JH1	6	2,936,936	33.0
		Streptococcus pyogenes M1 GAS (SF370)	6	1,852,441	38.5
		Bacillus subtilis W23	8	4,027,676	43.9
		Bacillus subtilis 168	10	4,215,606	43.5
		Brevibacillus brevis NBRC 100599	15	6,296,436	47.3
Proteobacteria[c]	2.2 ± 1.3 (α)	*Bartonella henselae* Houston-1	2	1,931,047	38.2
		Erythrobacter litoralis HTCC2594	1	3,052,398	63.1
	3.3 ± 1.6 (β)	Candidatus *Zinderia insecticola* CARI	1	208,564	13.5
	2.7 ± 1.4 (δ)	*Anaeromyxobacter dehalogenans* 2CP-C	2	5,013,479	74.9
		Desulfovibrio vulgaris Hildenborough	5	3,773,159	63.2
	3.0 ± 1.1 (ε)	*Helicobacter pylori* 26695	2	1,667,867	38.9
		Campylobacter jejuni 269.97	3	1,845,106	30.4
	5.8 ± 2.8 (γ)	*Buchnera aphidicola* (*Acyrthosiphon pisum*)	1	655,725	26.3
		Francisella tularensis FSC147	3	1,893,886	32.3
		Aggregatibacter actinomycetemcomitans D7S-1	6	2,309,073	44.3

(Continued)

Table 8.1 16S rRNA Gene Copy Numbers in a Subset of Bacterial Genomes (cont.)

Phylum	Average 16S rRNA Gene Copies in Phylum[a]	Organism Name	16S rRNA Gene Copy Number[b]	Genome Size (bp)[d,e]	GC%[e]
Proteobacteria[c]	5.8 ± 2.8 (γ)	*Haemophilus influenzae* 86-028NP	6	1,914,490	38.2
		Escherichia coli K-12 MG1655	7	4,641,652	50.8
		Yersinia pestis 91001	7	4,803,217	47.7
		Klebsiella pneumoniae HS11286	8	5,682,322	57.1
		Vibrio cholerae N16961	8	4,033,464	47.5
		Vibrio vulnificus MO6-24/O	9	5,007,768	47.0
		Aeromonas veronii B565	10	4,551,783	58.7
		Vibrio natriegens ATCC 14048	13	5,131,685	45.0
		Photobacterium profundum SS9	15	6,403,280	42.0
Spirochaetes	2.4 ± 1.0	*Borrelia burgdorferi* N40	1	1,339,539	28.6
		Treponema denticola ATCC 35405	2	2,843,201	37.9
		Treponema pallidum Chicago	2	1,139,281	52.8
Synergistetes	2.5 ± 1.0	*Anaerobaculum mobile* DSM 13181	2	2,160,700	48.0
		Thermanaerovibrio acidaminovorans DSM 6589	3	1,848,474	63.8
Tenericutes	1.6 ± 0.5	*Mycoplasma genitalium* G-37	1	580,076	31.7

[a]From Vetrovsky and Baldrian.[73]
[b]From the following sources: ribosomal RNA database (rrnDB).[74]
[c]Values are provided for each subdivisions. (α) *Alphaproteobacteria*, (β) *Betaproteobacteria*, (δ) *Deltaproteobacteria*, (ε) *Epsilonproteobacteria, and* (γ) *Gammaproteobacteria*.
[d]"bp" stands for base pairs.
[e]From National Center for Biotechnology Information (NCBI) Genome Information by Organism (www.ncbi.nlm.nih.gov/genome) and Kyoto Encyclopedia of Genes and Genomes (KEGG) Complete Genomes.[75]

the GC% range from 14% to 75%, while the number of ribosomal operon varies from 1 to 15. Thus, the interpretation of microbial diversity and abundances, (relative abundance distribution, estimate of abundance of different taxa, overall diversity, and similarity measurements) based on the phylogenetically informative 16S rRNA gene quantitation, should consider the variation in both the abundance of organisms and the operon copy numbers per genome. Refined analyses will only be available for a small community where all the partners are defined. A software is available that estimates both 16S rRNA gene copy number and abundance of organisms.[64] Further efforts need to be spent to relate these 16S rRNA gene copy number with genome copy number as discussed in the text below.

Not all bacteria conform to the patterns of genome organization, chromosomal replication initiation, elongation, termination, and genomic segregation during cell division exemplified by *Escherichia coli*, whose genome is distributed in one chromosome and has only one genome copy per cell. To be truly quantitative, we will also need to understand the ploidy of each organism in function of the experimental conditions (Table 8.2). The biological significance of polyploidy will depend on the system studied and might be involved in diverse functions such as DNA recombination among genome copies, replacement of deleterious mutations through homologous recombination of genomes, or to mitigate the accumulation of deleterious mutations over time.[65–69] Additionally, the cells can replicate asynchronously, displaying a heterogeneous DNA content.[70,71] We must contend with the fact that the genome copy number can change in the different phases of growth and that more than one ploidy can be observed in a population.[70,72] An understanding of the role of polyploidy and replication will provide insights into the extent the structure and content of the genome influences the phenotypic features of cells with multiple genomes, as well as influence the data from each "omics" platforms. In some remarkable cases there is a complementation of the physiology of both hosts and their polyploid symbionts, and these functional interactions remain to be elucidated.

Defining What Is a Strain

Bacteria, both in the laboratory and in nature, are studied at the population level. Bacterial populations are not composed of millions of identical individuals. During cell duplication, the genomes of individual cells are subjected to mutations, producing a genetically heterogeneous population within a species. Large-scale metagenomic studies reveal that microbial communities are predominantly organized in sequence-discrete populations, and the genomes of the organisms within those populations share higher than 94% average nucleotide identity (ANI). These sequence-discrete populations are important units within natural

Table 8.2 Genome Copy Numbers Per Cell of a Subset of Bacterial and Archaeal Species

Phylum	Organism Name	Genome Copy Number (Average or Range)	Ploidy	Generation Time, Growth Phase Environment-Free Living/Facultative-Obligate Symbiont	References
Bacteroidetes	*Blattabacterium sp.*	323–353 10–18	Polyploid Polyploid	Obligate endosymbiont of cockroach *Blattella oreintalis* Obligate endosymbiont of cockroach *Periplaneta americana*	Lopez-Sanchez *et al.*[76]
	Candidatus Sulcia muelleri DMIN	140–880	Polyploid	Obligate endosymbiont of green sharpshooter *Draeculecephala minerva*	Woyke *et al.*[77]
	Aphanizomenon ovalisporum	84–122 1–4	Polyploid Oligoploid	Akinetes (dormant spore-like cells) Vegetative cells	Sukenik *et al.*[78]
	Synechococcus PCC 7942	4	Oligoploid	Exponential and stationary phases (generation time 1440 min)	Griese *et al.*[67]
	Synechocystis PCC 6803 Motile wild-type	218 58 58	Polyploid	Exponential phase Linear phase (1200 min) Stationary phase	Griese *et al.*[67]
Deinococcus-Thermus	*Deinococcus radiodurans*	10 4–8	Oligoploid	Exponential phase Stationary phase	Hansen,[79] Minton[80]
	Thermus thermophiles HB8	4–5	Oligoploid	Exponential and stationary phase (slow growth conditions)	Ohtani *et al.*[81]
Firmicutes	*Epulopiscium sp.* Type B	20,000–400,000 49,000–120,000	Polyploid	Symbiont of the unicornfish *Naso tonganus* symbiont	Mendell *et al.*,[82] Angert[83]
	Lactobacillus lactis subsp. lactis IL1403	2	Diploid	Doubling time 223 min (slow growing culture)	Michelsen *et al.*,[84]
Proteobacteria	*Azotobacter vinelandii*	>40 >80 >100	Polyploid	Late exponential phase Early stationary phase Late stationary phase	Nagpal *et al.*,[85] Maldonado *et al.*[86]

Table 8.2 Genome Copy Numbers Per Cell of a Subset of Bacterial and Archaeal Species *(cont.)*

Phylum	Organism Name	Genome Copy Number (Average or Range)	Ploidy	Generation Time, Growth Phase Environment-Free Living/Facultative-Obligate Symbiont	References
Proteobacteria	*Buchnera sp.*	120 (50–200)	Polyploid	Obligate endosymbiont of the aphid *Acyrthosiphon pisum*; genome copy number varies with host developmental stage	Komaki and Ishikawa[69,87]
	Caulobacter crescentus	2.1	Monoploid	Doubling time 93 min	Pecoraro *et al.*[68]
	Desulfovibrio vulgaris	4	Oligoploid	Doubling time 2400 min	Postgate *et al.*[88]
	Escherichia coli	2.5/1.2[a] 6.8/1.7[a]	Monoploid Merooligoploid	Doubling time 103 min Doubling time 25 min	Pecoraro *et al.*[68]
	Neisseria gonorrhoeae	3	Oligoploid	Exponential phase (generation time 60 min)	Tobiason and Seifert[89]
	Pseudomonas putida	20/14[a]	Polyploid	Doubling time 46 min	Pecoraro *et al.*[68]
	Wolinella succinogenes	0.9	Monoploid	Doubling time 96 min	Pecoraro *et al.*[68]
Spirochaetes	*Borrelia hermsii*	5 14 (12–17)	Oligoploid Polyploid	Late exponential phase (maintained in laboratory) Isolated from mice	Kitten and Barbour[90]
Euryarchaeota	*Methanococcus maripaludis*	55 30	Polyploid	Exponential phase Stationary phase	Hildenbrand *et al.*[91]
	Methanothermobacter thermoautotrophicus	2 1–2	Diploid	Exponential phase Stationary phase	Majernik *et al.*[92]

[a]Based on gene copy number near origin/gene copy number near the termini.

microbial communities. Bacteria that belong to a particular population, but of different environment, significantly show less genetic identity to other co-occurring populations, typically less than 80–85% ANI. This genetic metric offers higher resolution than the widely used 16S rRNA gene sequencing analysis.[93,94] Defining strain might be contextual at first, until we have a more complete view of cell evolution. To facilitate the process, culture-independent "*omics*" techniques (transcriptomics, proteomics, and metabolomics) might further refine the taxonomical assignment and provide ecologically relevant properties of natural microbial populations. Quantification of yet-to-be-cultivated bacteria can be improved with the characterization of ecologically appropriate genes and pathways in sequence-discrete populations, which uniquely define the population genomic signatures.

Expanding Gene Catalogs

Identifying the genetic content of a microbiome is the first layer provided by the new generations of sequencing machines. From a metagenome or a metatranscriptome, the avalanche of information needs to be transformed in order to go beyond a simple comparison of gene counts. Genomic sequences from reference genomes are used in multiple aspects of the analysis, including gene definition, gene function, taxonomy, and so on. The first genome sequenced was isolated from the bacteria *Haemophilus influenza* in 1995.[95] Since then, the number of genome sequences has been growing rapidly and can be found in international depositories comprising DNA Data Bank of Japan (DDBJ), the European Nucleotide Archive (ENA), and the genetic sequence database of the National Center for Biotechnology Information (NCBI) of the United States (GenBank), as well as more specialized repositories. However, the number of reference genomes needs to increase to keep pace with advancements in metagenomics. Beyond cultivability, gene catalogs and single cell genomes will increase the pool of information to infer additional layers of analysis.[96–101]

Making Reference Strains Available

Presently, the number of cultivable strains deposited in reference strain depositories that are not yet sequenced is decreasing because of international efforts. The next frontier is in obtaining strains that were previously thought to be uncultivable. Some of the strains previously classified as "yet-to-be-cultivable" are now deposited at the American Type Culture Collection (ATCC) and sequenced by the means of the HMP,[99] awaiting further functional studies.

For a successful understanding of the microbiome and its interaction with the environment, novel large-scale investigations into the biology of single organisms and ecological models that integrate phylogenetic and functional relationships among organisms are required. Bacterial isolates available now or in the future will enable both biochemical-based study of their dynamic genomes and culture-based studies of their functional role in microbial communities. This will aid in improving assembly and annotation of metagenomes, and in quantification of microbial communities in their residing habitats.

Metabolic Potential

Bacteria exist in a wide range of environments and have extremely diverse physiological capabilities. Microbiome functionality can be derived either from gene-based knowledge or the intersection of other omics including metagenomics. Metabolism is key for the living cell. Databases such as KEGG, MetaCyc, Carbohydrate-Active enZYmes Database (CAZy), and Braunschweig Enzyme Database (BRENDA) considerably enhance our ability to create inferences leading to a greater understanding of single species or a complex community.[102–105] However, metabolism is not the only cell function, of which many aspects still remain unknown. For example, there is a large number of conserved proteins in international depositories for which a function needs to be identified to improve our understanding of the proteome.[106–108]

Learning About Archaea

Most previous work has focused on bacteria, as information about archaea is still nascent. Limited information is emerging about human–archaeal associations and the role of these organisms in human physiology. Much remains to be known about archaeal phylogenetic diversity, abundance, and biochemistry *in situ*. Current molecular approaches can reveal the genomic dynamics of methanogenic archaea associated with humans. These include *Methanobrevibacter smithii,* a methane producer predominant in human colon and also present in the vagina, *Methanobrevibacter oralis,* which has been associated with subgingival diseases and is capable to thrive at low pH in the stomach, and various other methanogens including *Methanosphaera stadtmanae*, *Methanobrevibacter millerae*, and *Methanobrevibacter arboriphilus.*[91,109–112] In the upcoming years, we need to expand our understanding of the role of archaea in the human microbiomes, as their transcripts are overabundant compared with their cell relative abundance.[113]

In closing, novel approaches are essential to properly integrate metagenomics, proteomics, lipidomics, and metabolomics in a comprehensive and integrative conceptual framework. Proper annotation of data sets is the first step in this direction by using minimum information standards when depositing the data sets and the annotations, and standardizing the names of body sites as well as of other descriptive components.[114–116] This opportunity is offered to all of us.

ACKNOWLEDGMENTS

We apologize for limiting the number of references on a vast topic. The laboratory is supported by the National Institutes of Health under Grant CA166150.

REFERENCES

1. Ringel-Kulka T, Cheng J, Ringel Y, Salojarvi J, Carroll I, Palva A, et al. Intestinal microbiota in healthy U.S. young children and adults – a high throughput microarray analysis. PloS One 2013;8(5):e64315.

2. Jost T, Lacroix C, Braegger CP, Chassard C. New insights in gut microbiota establishment in healthy breast fed neonates. PloS One 2012;7(8):e44595.

3. Trabal Fernandez N, Mazon-Suastegui JM, Vazquez-Juarez R, Ascencio-Valle F, Romero J. Changes in the composition and diversity of the bacterial microbiota associated with oysters (*Crassostrea corteziensis, Crassostrea gigas* and *Crassostrea sikamea*) during commercial production. FEMS Microbiol Ecol 2014;88(1):69–83.

4. Kohl KD, Cary TL, Karasov WH, Dearing MD. Restructuring of the amphibian gut microbiota through metamorphosis. Environ Microbiol Rep 2013;5(6):899–903.

5. Faith JJ, Guruge JL, Charbonneau M, Subramanian S, Seedorf H, Goodman AL, et al. The long-term stability of the human gut microbiota. Science 2013;341(6141):1237439.

6. Schloissnig S, Arumugam M, Sunagawa S, Mitreva M, Tap J, Zhu A, et al. Genomic variation landscape of the human gut microbiome. Nature 2013;493(7430):45–50.

7. Human Microbiome Project Consortium. Structure, function and diversity of the healthy human microbiome. Nature 2012;486(7402):207–14.

8. Segata N, Haake SK, Mannon P, Lemon KP, Waldron L, Gevers D, et al. Composition of the adult digestive tract bacterial microbiome based on seven mouth surfaces, tonsils, throat and stool samples. Genome Biol 2012;13(6):R42.

9. Holdeman LV, Good IJ, Moore WE. Human fecal flora: variation in bacterial composition within individuals and a possible effect of emotional stress. Appl Environ Microbiol 1976;31(3):359–75.

10. Tillisch K, Labus J, Kilpatrick L, Jiang Z, Stains J, Ebrat B, et al. Consumption of fermented milk product with probiotic modulates brain activity. Gastroenterology 2013;144(7):1394–401.

11. Round JL, Mazmanian SK. The gut microbiota shapes intestinal immune responses during health and disease. Nat Rev Immunol 2009;9(5):313–23.

12. Diaz Heijtz R, Wang S, Anuar F, Qian Y, Bjorkholm B, Samuelsson A, et al. Normal gut microbiota modulates brain development and behavior. Proc Natl Acad Sci USA 2011;108(7): 3047–52.

13. Adler CJ, Dobney K, Weyrich LS, Kaidonis J, Walker AW, Haak W, et al. Sequencing ancient calcified dental plaque shows changes in oral microbiota with dietary shifts of the Neolithic and Industrial revolutions. Nat Genet 2013;45(4):450–5.

14. Tito RY, Knights D, Metcalf J, Obregon-Tito AJ, Cleeland L, Najar F, et al. Insights from characterizing extinct human gut microbiomes. PloS One 2012;7(12):e51146.

15. Warinner C, Rodrigues JF, Vyas R, Trachsel C, Shved N, Grossmann J, et al. Pathogens and host immunity in the ancient human oral cavity. Nat Genet 2014;46(4):336–344.

16. Perez-Cobas AE, Artacho A, Knecht H, Ferrus ML, Friedrichs A, Ott SJ, et al. Differential effects of antibiotic therapy on the structure and function of human gut microbiota. PloS One 2013;8(11):e80201.

17. Dethlefsen L, Huse S, Sogin ML, Relman DA. The pervasive effects of an antibiotic on the human gut microbiota, as revealed by deep 16S rRNA sequencing. PLoS Biol 2008;6(11):e280.

18. Hill DA, Hoffmann C, Abt MC, Du Y, Kobuley D, Kirn TJ, et al. Metagenomic analyses reveal antibiotic-induced temporal and spatial changes in intestinal microbiota with associated alterations in immune cell homeostasis. Mucosal Immunol 2010;3(2):148–58.

19. Brace C, Gloor GB, Ropeleski M, Allen-Vercoe E, Petrof EO. Microbial composition analysis of *Clostridium difficile* infections in an ulcerative colitis patient treated with multiple fecal microbiota transplantations. J Crohn's Colitis 2014;8(9):1133–1137.

20. Song Y, Garg S, Girotra M, Maddox C, von Rosenvinge EC, Dutta A, et al. Microbiota dynamics in patients treated with fecal microbiota transplantation for recurrent *Clostridium difficile* infection. PloS One 2013;8(11):e81330.

21. Viaud S, Saccheri F, Mignot G, Yamazaki T, Daillere R, Hannani D, et al. The intestinal microbiota modulates the anticancer immune effects of cyclophosphamide. Science 2013; 342(6161):971–6.

22. Goleva E, Jackson LP, Harris JK, Robertson CE, Sutherland ER, Hall CF, et al. The effects of airway microbiome on corticosteroid responsiveness in asthma. Am J Respir Crit Care Med 2013;188(10):1193–201.

23. Jiang B, Liang X, Chen Y, Ma T, Liu L, Li J, et al. Integrating next-generation sequencing and traditional tongue diagnosis to determine tongue coating microbiome. Sci Rep 2012;2:936.

24. Zhu L, Wu Q, Dai J, Zhang S, Wei F. Evidence of cellulose metabolism by the giant panda gut microbiome. Proc Natl Acad Sci USA 2011;108(43):17714–9.

25. Turnbaugh PJ, Hamady M, Yatsunenko T, Cantarel BL, Duncan A, Ley RE, et al. A core gut microbiome in obese and lean twins. Nature 2009;457(7228):480–4.

26. Amato KR, Yeoman CJ, Kent A, Righini N, Carbonero F, Estrada A, et al. Habitat degradation impacts black howler monkey (*Alouatta pigra*) gastrointestinal microbiomes. ISME J 2013;7(7):1344–53.

27. David LA, Maurice CF, Carmody RN, Gootenberg DB, Button JE, Wolfe BE, et al. Diet rapidly and reproducibly alters the human gut microbiome. Nature 2014;505(7484):559–63.

28. Lavery TJ, Roudnew B, Seymour J, Mitchell JG, Jeffries T. High nutrient transport and cycling potential revealed in the microbial metagenome of Australian sea lion (*Neophoca cinerea*) faeces. PloS One 2012;7(5):e36478.

29. Ridaura VK, Faith JJ, Rey FE, Cheng J, Duncan AE, Kau AL, et al. Gut microbiota from twins discordant for obesity modulate metabolism in mice. Science 2013;341(6150): 1241214.

30. Jung JY, Lee SH, Kim JM, Park MS, Bae JW, Hahn Y, et al. Metagenomic analysis of kimchi, a traditional Korean fermented food. Appl Environ Microbiol 2011;77(7):2264–74.

31. De Filippis F, La Storia A, Stellato G, Gatti M, Ercolini D. A selected core microbiome drives the early stages of three popular italian cheese manufactures. PloS One 2014;9(2):e89680.

32. Badri DV, Chaparro JM, Zhang R, Shen Q, Vivanco JM. Application of natural blends of phytochemicals derived from the root exudates of Arabidopsis to the soil reveal that phenolic-related compounds predominantly modulate the soil microbiome. J Biol Chem 2013;288(7):4502–12.

33. Larsen AM, Mohammed HH, Arias CR. Characterization of the gut microbiota of three commercially valuable warmwater fish species. J Appl Microbiol 2014;116(6):1396–1404.

34. Hwang C, Ling F, Andersen GL, LeChevallier MW, Liu WT. Microbial community dynamics of an urban drinking water distribution system subjected to phases of chloramination and chlorination treatments. App Environ Microbiol 2012;78(22):7856–65.

35. Hong PY, Hwang C, Ling F, Andersen GL, LeChevallier MW, Liu WT. Pyrosequencing analysis of bacterial biofilm communities in water meters of a drinking water distribution system. App Environ Microbiol 2010;76(16):5631–5.

36. Holinger EP, Ross KA, Robertson CE, Stevens MJ, Harris JK, Pace NR. Molecular analysis of point-of-use municipal drinking water microbiology. Water Res 2014;49:225–35.

37. Kembel SW, Jones E, Kline J, Northcutt D, Stenson J, Womack AM, et al. Architectural design influences the diversity and structure of the built environment microbiome. ISME J 2012;6(8):1469–79.

38. Meadow JF, Altrichter AE, Kembel SW, Moriyama M, O'Connor TK, Womack AM, et al. Bacterial communities on classroom surfaces vary with human contact. Microbiome 2014;2(1):7.

39. Stephenson RE, Gutierrez D, Peters C, Nichols M, Boles BR. Elucidation of bacteria found in car interiors and strategies to reduce the presence of potential pathogens. Biofouling 2014;30(3):337–46.

40. Robertson CE, Baumgartner LK, Harris JK, Peterson KL, Stevens MJ, Frank DN, et al. Culture-independent analysis of aerosol microbiology in a metropolitan subway system. App Environ Microbiol 2013;79(11):3485–93.

41. DeLeon-Rodriguez N, Lathem TL, Rodriguez RL, Barazesh JM, Anderson BE, Beyersdorf AJ, et al. Microbiome of the upper troposphere: species composition and prevalence, effects of tropical storms, and atmospheric implications. Proc Natl Acad Sci USA 2013;110(7):2575–80.

42. Costello V. Students Aged 9 to 65+ Study PLOS Research in Marine Megafauna MOOC. Web blog post Diverse perspectives on science and medicine PLOS Blogs 2014, 4 April 2014.

43. Ho AD, Reich J, Nesterko SO, Seaton DT, Mullaney T, Waldo J, et al. HarvardX and MITx: The First Year of Open Online Courses, Fall 2012-Summer 2013 Social Science Research Network SSRN Electronic Journal pp: 33 2014.

44. Crall AW, Jordan R, Holfelder K, Newman GJ, Graham J, Waller DM. The impacts of an invasive species citizen science training program on participant attitudes, behavior, and science literacy. Public Understand Sci 2013;22(6):745–64.

45. Janssens AC, Kraft P. Research conducted using data obtained through online communities: ethical implications of methodological limitations. PLoS Med 2012;9(10):e1001328.

46. Davis PM, Walters WH. The impact of free access to the scientific literature: a review of recent research. J Med Lib Assoc 2011;99(3):208–17.

47. Venter JC, Remington K, Heidelberg JF, Halpern AL, Rusch D, Eisen JA, et al. Environmental genome shotgun sequencing of the Sargasso Sea. Science 2004;304(5667):66–74.

48. Zhou Y, Gao H, Mihindukulasuriya KA, Rosa PS, Wylie KM, Vishnivetskaya T, et al. Biogeography of the ecosystems of the healthy human body. Genome Biol 2013;14(1):R1.

49. Kolenbrander PE, Palmer RJ, Periasamy S, Jakubovics NS. Oral multispecies biofilm development and the key role of cell-cell distance. Nat Rev Micro 2010;8(7):471–80.

50. Faust K, Sathirapongsasuti JF, Izard J, Segata N, Gevers D, Raes J, et al. Microbial Co-occurrence Relationships in the Human Microbiome. PLoS Computat Biol 2012;8(7):e1002606.

51. Smillie CS, Smith MB, Friedman J, Cordero OX, David LA, Alm EJ. Ecology drives a global network of gene exchange connecting the human microbiome. Nature 2011;480(7376):241–4.

52. Hehemann JH, Correc G, Barbeyron T, Helbert W, Czjzek M, Michel G. Transfer of carbohydrate-active enzymes from marine bacteria to Japanese gut microbiota. Nature 2010;464(7290):908–12.

53. Ramsey MM, Whiteley M. Polymicrobial interactions stimulate resistance to host innate immunity through metabolite perception. Proc Natl Acad Sci USA 2009;106(5):1578–83.

54. Sieber JR, McInerney MJ, Gunsalus RP. Genomic insights into syntrophy: the paradigm for anaerobic metabolic cooperation. Annu Rev Microbiol 2012;66:429–52.

55. Qin J, Li R, Raes J, Arumugam M, Burgdorf KS, Manichanh C, et al. A human gut microbial gene catalogue established by metagenomic sequencing. Nature 2010;464(7285):59–65.

56. Stackebrandt E. Forces shaping bacterial systematics. Microbe 2007;2(6):283–8.

57. Nelson KE, Weinstock GM, Highlander SK, Worley KC, Creasy HH, Wortman JR, et al. A catalog of reference genomes from the human microbiome. Science 2010;328(5981):994–9.

58. Dewhirst FE, Chen T, Izard J, Paster BJ, Tanner AC, Yu WH, et al. The human oral microbiome. J Bacteriol 2010;192(19):5002–17.

59. Huang K, Brady A, Mahurkar A, White O, Gevers D, Huttenhower C, et al. MetaRef: a pan-genomic database for comparative and community microbial genomics. Nucleic Acids Res 2014;42(1):D617–24.

60. Chen T, Yu WH, Izard J, Baranova OV, Lakshmanan A, Dewhirst FE. The Human Oral Microbiome Database: a web accessible resource for investigating oral microbe taxonomic and genomic information. Database: J Biol Databases Curat 2010;2010. baq013.

61. Cole JR, Wang Q, Fish JA, Chai B, McGarrell DM, Sun Y, et al. Ribosomal Database Project: data and tools for high throughput rRNA analysis. Nucleic Acids Res 2014;42(Database issue): D633–642.

62. McDonald D, Price MN, Goodrich J, Nawrocki EP, DeSantis TZ, Probst A, et al. An improved Greengenes taxonomy with explicit ranks for ecological and evolutionary analyses of bacteria and archaea. ISME J 2012;6(3):610–8.

63. Quast C, Pruesse E, Yilmaz P, Gerken J, Schweer T, Yarza P, et al. The SILVA ribosomal RNA gene database project: improved data processing and web-based tools. Nucleic Acids Res 2013;41(Database issue):D590–6.

64. Kembel SW, Wu M, Eisen JA, Green JL. Incorporating 16S gene copy number information improves estimates of microbial diversity and abundance. PLoS Computat Biol 2012;8(10):e1002743.

65. Slade D, Lindner AB, Paul G, Radman M. Recombination and replication in DNA repair of heavily irradiated *Deinococcus radiodurans*. Cell 2009;136(6):1044–55.

66. Zahradka K, Slade D, Bailone A, Sommer S, Averbeck D, Petranovic M, et al. Reassembly of shattered chromosomes in *Deinococcus radiodurans*. Nature 2006;443(7111):569–73.

67. Griese M, Lange C, Soppa J. Ploidy in cyanobacteria. FEMS Microbiol Lett 2011;323(2): 124–31.

68. Pecoraro V, Zerulla K, Lange C, Soppa J. Quantification of ploidy in proteobacteria revealed the existence of monoploid, (mero-)oligoploid and polyploid species. PloS One 2011;6(1):e16392.

69. Komaki K, Ishikawa H. Genomic copy number of intracellular bacterial symbionts of aphids varies in response to developmental stage and morph of their host. Insect Biochem Mol Biol 2000;30(3):253–8.

70. Muller S, Babel W. Analysis of bacterial DNA patterns – an approach for controlling biotechnological processes. J Microbiol Methods 2003;55(3):851–8.

71. Muller S, Nebe-von-Caron G. Functional single-cell analyses: flow cytometry and cell sorting of microbial populations and communities. FEMS Microbiol Rev 2010;34(4):554–87.

72. Caro A, Gros O, Got P, De Wit R, Troussellier M. Characterization of the population of the sulfur-oxidizing symbiont of *Codakia orbicularis* (Bivalvia, Lucinidae) by single-cell analyses. Appl Environ Microbiol 2007;73(7):2101–9.

73. Vetrovsky T, Baldrian P. The variability of the 16S rRNA gene in bacterial genomes and its consequences for bacterial community analyses. PloS One 2013;8(2):e57923.

74. Lee ZM, Bussema C 3rd, Schmidt TM. rrnDB: documenting the number of rRNA and tRNA genes in bacteria and archaea. Nucleic Acids Res 2009;37(Database issue):D489–93.

75. Kanehisa M, Araki M, Goto S, Hattori M, Hirakawa M, Itoh M, et al. KEGG for linking genomes to life and the environment. Nucleic Acids Res 2008;36(Database issue):D480–4.

76. Lopez-Sanchez MJ, Neef A, Patino-Navarrete R, Navarro L, Jimenez R, Latorre A, et al. Blattabacteria, the endosymbionts of cockroaches, have small genome sizes and high genome copy numbers. Environ Microbiol 2008;10(12):3417–22.

77. Woyke T, Tighe D, Mavromatis K, Clum A, Copeland A, Schackwitz W, et al. One bacterial cell, one complete genome. PloS One 2010;5(4):e10314.

78. Sukenik A, Kaplan-Levy RN, Welch JM, Post AF. Massive multiplication of genome and ribosomes in dormant cells (akinetes) of Aphanizomenon ovalisporum (Cyanobacteria). ISME J 2012;6(3):670–9.

79. Hansen MT. Multiplicity of genome equivalents in the radiation-resistant bacterium *Micrococcus radiodurans*. J Bacteriol 1978;134(1):71–5.

80. Minton KW. DNA repair in the extremely radioresistant bacterium *Deinococcus radiodurans*. Mol Microbiol 1994;13(1):9–15.

81. Ohtani N, Tomita M, Itaya M. An extreme thermophile, *Thermus thermophilus*, is a polyploid bacterium. J Bacteriol 2010;192(20):5499–505.

82. Mendell JE, Clements KD, Choat JH, Angert ER. Extreme polyploidy in a large bacterium. Proc Natl Acad Sci USA 2008;105(18):6730–4.

83. Angert ER. DNA replication and genomic architecture of very large bacteria. Annu Rev Microbiol 2012;66:197–212.

84. Michelsen O, Hansen FG, Albrechtsen B, Jensen PR. The MG1363 and IL1403 laboratory strains of *Lactococcus lactis* and several dairy strains are diploid. J Bacteriol 2010;192(4):1058–65.

85. Nagpal P, Jafri S, Reddy MA, Das HK. Multiple chromosomes of *Azotobacter vinelandii*. J Bacteriol 1989;171(6):3133–8.

86. Maldonado R, Jimenez J, Casadesus J. Changes of ploidy during the *Azotobacter vinelandii* growth cycle. J Bacteriol 1994;176(13):3911–9.

87. Komaki K, Ishikawa H. Intracellular bacterial symbionts of aphids possess many genomic copies per bacterium. J Mol Evol 1999;48(6):717–22.

88. Postgate JR, Kent HM, Robson RL, Chesshyre JA. The genomes of *Desulfovibrio gigas* and *D. vulgaris*. J Gen Microbiol 1984;130(7):1597–601.

89. Tobiason DM, Seifert HS. The obligate human pathogen, *Neisseria gonorrhoeae*, is polyploid. PLoS Biol 2006;4(6):e185.

90. Kitten T, Barbour AG. The relapsing fever agent *Borrelia hermsii* has multiple copies of its chromosome and linear plasmids. Genetics 1992;132(2):311–24.

91. Hildenbrand C, Stock T, Lange C, Rother M, Soppa J. Genome copy numbers and gene conversion in methanogenic archaea. J Bacteriol 2011;193(3):734–43.

92. Majernik AI, Lundgren M, McDermott P, Bernander R, Chong JP. DNA content and nucleoid distribution in *Methanothermobacter thermautotrophicus*. J Bacteriol 2005;187(5):1856–8.

93. Goris J, Konstantinidis KT, Klappenbach JA, Coenye T, Vandamme P, Tiedje JM. DNA-DNA hybridization values and their relationship to whole-genome sequence similarities. Int J Syst Evol Microbiol 2007;57(Pt 1):81–91.

94. Caro-Quintero A, Konstantinidis KT. Bacterial species may exist, metagenomics reveal. Environ Microbiol 2012;14(2):347–55.

95. Fleischmann RD, Adams MD, White O, Clayton RA, Kirkness EF, Kerlavage AR, et al. Whole-genome random sequencing and assembly of *Haemophilus influenzae* Rd. Science 1995;269(5223):496–512.

96. Kyrpides NC, Hugenholtz P, Eisen JA, Woyke T, Goker M, Parker CT, et al. Genomic encyclopedia of bacteria and archaea: sequencing a myriad of type strains. PLoS Biol 2014;12(8):e1001920.

97. Beall CJ, Campbell AG, Dayeh DM, Griffen AL, Podar M, Leys EJ. Single cell genomics of uncultured, health-associated *Tannerella* BU063 (Oral Taxon 286) and comparison to the closely related pathogen *Tannerella forsythia*. PloS One 2014;9(2):e89398.

98. Mason OU, Hazen TC, Borglin S, Chain PS, Dubinsky EA, Fortney JL, et al. Metagenome, metatranscriptome and single-cell sequencing reveal microbial response to deepwater horizon oil spill. ISME J 2012;6(9):1715–27.

99. Human Microbiome Project Consortium. A framework for human microbiome research. Nature 2012;486(7402):215–21.

100. Qin J, Li R, Raes J, Arumugam M, Burgdorf KS, Manichanh C, et al. A human gut microbial gene catalogue established by metagenomic sequencing. Nature 2010;464(7285):59–65.

101. Rinke C, Schwientek P, Sczyrba A, Ivanova NN, Anderson IJ, Cheng JF, et al. Insights into the phylogeny and coding potential of microbial dark matter. Nature 2013;499(7459):431–7.

102. Lombard V, Golaconda Ramulu H, Drula E, Coutinho PM, Henrissat B. The carbohydrate-active enzymes database (CAZy) in 2013. Nucleic Acids Res 2014;42(1):D490–5.

103. Caspi R, Altman T, Dale JM, Dreher K, Fulcher CA, Gilham F, et al. The MetaCyc database of metabolic pathways and enzymes and the BioCyc collection of pathway/genome databases. Nucleic Acids Res 2010;38(Database issue):D473–9.

104. Kanehisa M, Goto S, Furumichi M, Tanabe M, Hirakawa M. KEGG for representation and analysis of molecular networks involving diseases and drugs. Nucleic Acids Res 2010;38 (Database issue):D355–60.

105. Schomburg I, Chang A, Placzek S, Sohngen C, Rother M, Lang M, et al. BRENDA in 2013 integrated reactions, kinetic data, enzyme function data, improved disease classification: new options and contents in BRENDA. Nucleic Acids Res 2013;41(Database issue):D764–72.

106. Wilke A, Harrison T, Wilkening J, Field D, Glass EM, Kyrpides N, et al. The M5nr: a novel non-redundant database containing protein sequences and annotations from multiple sources and associated tools. BMC Bioinform 2012;13:141.

107. UniProt C. Activities at the Universal Protein Resource (UniProt). Nucleic Acids Res 2014;42(Database issue):D191–8.

108. Goodacre NF, Gerloff DL, Uetz P. Protein domains of unknown function are essential in bacteria. mBio 2013;5(1):e00744–e813.

109. Samuel BS, Hansen EE, Manchester JK, Coutinho PM, Henrissat B, Fulton R, et al. Genomic and metabolic adaptations of *Methanobrevibacter smithii* to the human gut. Proc Natl Acad Sci USA 2007;104(25):10643–8.

110. Dridi B, Raoult D, Drancourt M. Archaea as emerging organisms in complex human microbiomes. Anaerobe 2011;17(2):56–63.

111. Aminov RI. Role of archaea in human disease. Front Cell Infect Microbiol 2013;3:42.

112. Khelaifia S, Garibal M, Robert C, Raoult D, Drancourt M. Draft genome sequence of a human-associated isolate of *Methanobrevibacter arboriphilicus*, the Lowest-G+C-content archaeon. Genome Announce 2014;2(1):e01181-13.

113. Franzosa EA, Morgan XC, Segata N, Waldron L, Reyes J, Earl AM, et al. Relating the metatranscriptome and metagenome of the human gut. Proc Natl Acad Sci USA 2014;111(22): E2329–2338.

114. Yilmaz P, Kottmann R, Field D, Knight R, Cole JR, Amaral-Zettler L, et al. Minimum information about a marker gene sequence (MIMARKS) and minimum information about any (x) sequence (MIxS) specifications. Nat Biotechnol 2011;29(5):415–20.

115. Field D, Amaral-Zettler L, Cochrane G, Cole JR, Dawyndt P, Garrity GM, et al. The Genomic Standards Consortium. PLoS Biol 2011;9(6):e1001088.

116. Field D, Garrity G, Gray T, Morrison N, Selengut J, Sterk P, et al. The minimum information about a genome sequence (MIGS) specification. Nat Biotechnol 2008;26(5):541–7.

CHAPTER 9

Toward the Understanding of the Human Virome

Nadim J. Ajami and Joseph F. Petrosino

INTRODUCTION

The collection of viral genomes present in any given sample is referred to as the viral metagenome or virome. The Earth virome is estimated to be composed of 10^{31} viral particles associated or not with a host, including humans.[1–3] The study and our understanding of viromes has been stimulated by the advent of next-generation sequencing platforms accompanied by improvements in unbiased viral nucleic acid extraction methods and decreasing sequencing costs. The virome contains the most abundant and fastest mutating genetic elements of the Earth.[4] Viruses interact with the human host through its bacterial communities (prokaryotic viruses) as well as by residing or interacting with human cells in acute, persistent, or latent infections.[4] The genetic characterization of the virome, as a constituent of the microbiome, has trailed behind analyses of the bacterial microbiome, in part, because of the lack of a conserved sequence that could be readily used to assign taxonomy and because many of the genes encoded in the virome have not been previously annotated.

The genomic age began in 1977 when the *Escherichia coli* bacteriophage ΦX174 was sequenced.[6] Twenty-five years later, in 2002, the viral metagenomics area arose with the publication of two uncultured marine viral communities.[7] Viral metagenomics has revolutionized the field of viral ecology by providing an unbiased, culture-independent, and high-throughput method to study the structure, function, and metabolic potential of viral communities and their environmental impact. Unlike traditional laboratory techniques for microbial and viral identification, metagenomics does not require prior isolation and clonal culturing for species characterization. The use of metagenomics has been particularly suitable for providing a general overview of the community structure

Metagenomics for Microbiology. http://dx.doi.org/10.1016/B978-0-12-410472-3.00009-9

(richness and abundance) and its functional potential (gene products). In principle, it allows the identification of any organism, including those commonly not detected because they are difficult to isolate and grow under laboratory conditions. Such organisms are estimated to constitute between 90% and 99% of microbial species.[8,9]

The use of viral metagenomics can be applied to a wide variety of fields ranging from ecology to environmental sciences,[7,10,11] the chemical industry,[12] and human health.[13–18] The most powerful use of viral metagenomics is in its application to the characterization of viral communities.[3,13,19–22] Breitbart performed the first example of viral metagenomics in 2002 where she and her colleagues revealed that viral diversity had been widely underestimated because in approximately 200 liters of marine water, more than 7000 different viral genomes were found.[7] This high degree of viral genetic diversity has been confirmed by further metagenomic studies of marine samples.[23–25]

VIROME CHARACTERIZATION WORKFLOW

The current data workflow involving metagenomic analyses initiates with quality control and preprocessing of raw reads produced by high-throughput sequencing technologies with the goal of creating a high-quality metagenomic dataset truly representative of the genotypes and their relative abundance in a sample. Quality control measures include the investigation of length, quality scores associated with each base, GC content, number of ambiguous bases, and the sequence complexity. All of these parameters are dependent on the sequencing technology and the upstream processes involved. Subsequently, after initial quality control assessment, foreign (nonviral) sequences are filtered resulting in a viral-only dataset. This filtering has to be carefully evaluated as part of these sequence data might come from genes of bacterial origin transferred to phages or from erroneously annotated sequences.[26,27] The taxonomic characterization of the virome relies on similarity-based methods usually employing BLAST searches,[28] although other useful algorithms exist such as FAAST, BLAT, VIROME, and MEGAN.[29–31] Searches based on stringent *E*-values can yield too few classifiable sequences and, in contrast, less-stringent *E*-values can result in a high number of incorrect assignments.[13] Moreover, the content of public viral sequence databases is incomplete and poorly reflects only the existing biological diversity.[19] In addition, viruses exhibit a high genetic diversity and divergence, which, in turn, limits the probability of finding similarities based on nucleic acid alignments. As an alternative, translated nucleotide sequences are used to aid on the classification of viral sequences because synonymous mutations are bypassed in the translation step.[32]

RECENT VIROME STUDIES

The collection of viruses found in humans includes viral particles capable of infecting eukaryotic, bacterial (bacteriophages), archaeal cells, and virus-derived genetic elements inserted within host chromosomes that have the potential to generate infectious particles, express proteins, and alter host–gene expression [proviruses, prophages, endogenous retroviruses (ERVs), and endogenous viral elements (EVEs)].[4] Characterization of the human virome has been mostly attempted using fecal samples.[14,33–38] Initial studies sought after DNA viruses in a stool sample obtained from a healthy individual and results showed most of the sequences generated were unknown. Among the identifiable viral sequences, the majority corresponded to bacteriophages and the community was estimated to have a high richness (~1200 genotypes) and diversity.[34] Similarly to previous findings, Breitbart *et al.*[33] reported an elevated percentage of unknown sequences (66%) and a significant abundance of phages in a study using feces of a 1-week-old infant.[33] Comparable observations were also reported by two studies on the DNA virome of the human gut, in which the percentages of unknown sequences were 81% and 98%, respectively, and phages dominated the viral community.[14,37]

INTERACTION OF THE VIROME WITH THE HUMAN HOST

In addition to DNA viruses, RNA viruses of the human gut have also been studied. In a study performed using stool samples form healthy adults, Zhang *et al.*[38] found that only 8.9% of the sequences were unknown and that among the identifiable viral sequences there were an insignificant number of phages. The majority of the identifiable viruses were plant viruses (91.5%), presumably introduced through consumption of contaminated produces.[38]

In humans, as in other mammals, viral genomes can replicate and persist in most nucleated cells. Viruses can be detected after clinical infections are resolved, usually hidden from the immune system in cells including neurons, hematopoietic cells, stem cells, and vascular endothelial cells.

Analogous to the bacterial microbiome, the virome is also in constant interaction with the host immune system maintaining a dynamic equilibrium. Constant exposure to low-virulence viruses and subclinical infection stimulate mucosal immune responses eliciting activation of immune cells and conferring resistance to other infections.[39] It is estimated that an individual healthy human harbors over 10 permanent chronic systemic viral infections; however, the chronic carriage of viruses is likely to be underestimated because of the lack of broadly used inexpensive techniques to detect and quantify extremely diverse and scarce members of the virome.[4,40] Nonetheless, viruses including herpesviruses, polyomaviruses, anellovirues, adenoviruses, and papillomaviruses have been shown to continuously activate the immune system through responses to pathogen-associated molecular patterns, and antigens are generated as these viruses reactivate from latency or continually replicate.[4]

Retroviruses are the only known eukaryotic viruses that require chromosomal integration as part of their replication cycle. However, elements from multiple other virus species have been identified to be integrated.[41,42] The presence of EVEs in our genomes suggest a role of viruses in gene transfer and evolution.[41] ERVs comprise 8% of the human genome and they are the largest contributors to the EVEs found in our genomes.[42,43] The impact of EVEs in host biology include the introduction of genetic variation, regulation of host gene expression, and coding of viral components, all of which have the potential to influence the host immune to self and foreign antigens.[42]

Asymptomatic systemic viral infections can impact the host by playing an important role in regulating the transcriptional state of healthy people and have been linked to diseases such as asthma and type 1 diabetes.[39] Constant but low levels of immune activation often result in inflammatory states and idiopathic systemic inflammation has been linked to pathologies including diabetes, cardiovascular diseases, and metabolic syndrome. Although the etiologies of this inflammation are not fully understood, the systemic and mucosal virome is sought to play a role.[4] In addition, the systemic virus may also contribute to the diversity of phenotypes observed in hosts upon other infections.

Many studies centered on the bacterial component of the microbiome attribute results to alterations in the bacterial community without acknowledging the virome and its relation to other components of the human microbiome.[44] There is extensive literature describing interactions between virus and bacteria in respiratory tract infections. This is exemplified by the increase in pneumococcal and staphylococcal pneumonia seen during influenza pandemics. Furthermore, viruses have been associated with increased severity of bacterial infections of the respiratory system.[39,45] These trans-kingdom interactions are not uncommon in mammals, since recent studies suggest that bacteria contribute to the infectivity of retroviruses and enteroviruses by direct interaction of viruses with bacterial products.[46–48]

Viruses that infect bacteria are highly variable and diverse across individuals.[36,37] Bacteriophages directly impact the structure of the bacterial microbiome via gene transfer and lysis of bacterial cells, conferring new phenotypes and opening niches allowing invasion or overgrowth of other bacteria, respectively.[49] Bacteriophages can have an impact in the host immune system by modifying bacterial cells and triggering inflammatory responses through toll-like receptor signaling. Such effects have also been observed in parasitic infections with *Leishmania* and *Trichomonas.*[50,51] It has been proposed that bacteriophages can interact with the human host by direct contact with epithelial cells and accessing the lamina propria through breaks in the intestinal mucosa subsequently spreading systemically.[49,52] Bacteriophage capsid antigens are also immunogenic and can elicit antibody responses in human[53] in addition to stimulating the production of cytokines such as interleukin 1β and tumor necrosis factor-α by macrophages.[54]

Analogously, the interaction of the members of the virome with other components of the microbiome can dictate the outcome of a disease or a treatment. As explained above, the biological effects of the virome extends through its interaction with the microbiome and the host components. This is also applicable to experimental animal models of human diseases[55–57] and needs to be taken in consideration when testing hypothesis and determining treatment efficacies. As an example, mice chronically infected with herpesvirus can increase resistance to *Listeria monocytogenes* and *Yersinia pestis* infection,[58] activate NK cells and increase resistance to tumor grafts,[59] delay onset of type 1 diabetes in non-obese diabetic mice,[60] and decrease adenoviral infection.[61] Nonetheless, the same infections can increase susceptibility to autoantigen-driven experimental allergic encephalomyelitis,[62] and affect malarial lethality.[63] Consequently, the complexity of the virome–host through direct and indirect interaction is likely to be high, thus having an unapparent effect in animal models of human disease.[4]

CONCLUDING REMARKS

Viral metagenomics analyses have shown that more than 60% of the sequences in a viral preparation are unique, thus representing unknown viral species.[19] Our current capacity of characterizing the members of the virome relies almost exclusively on the similarity of identified nucleic acid and protein sequences with published information. A major obstacle in defining a virus is using nucleic acid sequence alignments given the enormous variability of these organisms at this level. Therefore, it is very likely that the virome includes novel viruses that have yet to be characterized.

The dominant approach to the development of therapeutics is based on the fact that diseases are caused by single etiologies and interaction of these etiologies with their host is what is usually targeted. Although this approach has proven to be successful in some cases (e.g., antibiotics and vaccines), it often obviates the role of the metagenome, its components, and its crosstalk among members and the host.

Understanding the complexity of the effects of the virome in the host including pathogenic and nonpathogenic states will require efforts beyond the identification and characterization of viral etiologies. As its interaction with other members of the human microbiome and with the host during subclinical infections is further unraveled, we will be able to determine its true impact in human health. These efforts will undeniably lead to rapid clinical diagnosis, modulation of clinical therapies, and genome mining with industrial applications.[5]

REFERENCES

1. Rohwer F. Global phage diversity. Cell 2003;113(2):141.
2. Schoenfeld T, Liles M, Wommack KE, Polson SW, Godiska R, Mead D. Functional viral metagenomics and the next generation of molecular tools. Trends Microbiol 2010;18(1):20–9.

3. Suttle CA. Viruses in the sea. Nature 2005;437(7057):356–61.
4. Virgin HW. The virome in mammalian physiology and disease. Cell 2014;157(1):142–50.
5. Petrosino JF, Highlander S, Luna RA, Gibbs RA, Versalovic J. Metagenomic pyrosequencing and microbial identification. Clin Chem 2009;55(5):856–66.
6. Sanger F, Air GM, Barrell BG, Brown NL, Coulson AR, Fiddes CA, et al. Nucleotide sequence of bacteriophage phi X174 DNA. Nature 1977;265(5596):687–95.
7. Breitbart M, Salamon P, Andresen B, Mahaffy JM, Segall AM, Mead D, et al. Genomic analysis of uncultured marine viral communities. Proc Natl Acad Sci USA 2002;99(22):14250–5.
8. Rappe MS, Giovannoni SJ. The uncultured microbial majority. Annu Rev Microbiol 2003;57:369–94.
9. Pace NR. A molecular view of microbial diversity and the biosphere. Science 1997;276(5313):734–40.
10. Dinsdale EA, Rohwer F. Dissecting microbial employment. Nat Biotechnol 2008;26(9):997–8.
11. Dinsdale EA, Edwards RA, Hall D, Angly F, Breitbart M, Brulc JM, et al. Functional metagenomic profiling of nine biomes. Nature 2008;452(7187):629–32.
12. Lorenz P, Eck J. Metagenomics and industrial applications. Nat Rev Microbiol 2005;3(6):510–6.
13. Fancello L, Raoult D, Desnues C. Computational tools for viral metagenomics and their application in clinical research. Virology 2012;434(2):162–74.
14. Minot S, Sinha R, Chen J, Li H, Keilbaugh SA, Wu GD, et al. The human gut virome: inter-individual variation and dynamic response to diet. Genome Res 2011;21(10):1616–25.
15. Nakamura S, Yang CS, Sakon N, Ueda M, Tougan T, Yamashita A, et al. Direct metagenomic detection of viral pathogens in nasal and fecal specimens using an unbiased high-throughput sequencing approach. PloS One 2009;4(1):e4219.
16. Ravel J, Gajer P, Abdo Z, Schneider GM, Koenig SS, McCulle SL, et al. Vaginal microbiome of reproductive-age women. Proc Natl Acad Sci USA 2011;108(Suppl 1):4680–7.
17. Sullivan PF, Allander T, Lysholm F, Goh S, Persson B, Jacks A, et al. An unbiased metagenomic search for infectious agents using monozygotic twins discordant for chronic fatigue. BMC Microbiol 2011;11:2.
18. Turnbaugh PJ, Ley RE, Hamady M, Fraser-Liggett CM, Knight R, Gordon JI. The human microbiome project. Nature 2007;449(7164):804–10.
19. Edwards RA, Rohwer F. Viral metagenomics. Nat Rev Microbiol 2005;3(6):504–10.
20. Ma Y, Madupu R, Karaoz U, Nossa CW, Yang L, Yooseph S, et al. Human papillomavirus community in healthy persons, defined by metagenomics analysis of human microbiome project shotgun sequencing data sets. J Virol 2014;88(9):4786–97.
21. Minot S, Bryson A, Chehoud C, Wu GD, Lewis JD, Bushman FD. Rapid evolution of the human gut virome. Proc Natl Acad Sci USA 2013;110(30):12450–5.
22. Wylie KM, Weinstock GM, Storch GA. Emerging view of the human virome. Transl Res: J Lab Clin Med 2012;160(4):283–90.
23. Angly FE, Felts B, Breitbart M, Salamon P, Edwards RA, Carlson C, et al. The marine viromes of four oceanic regions. PLoS Biol 2006;4(11):e368.
24. Breitbart M, Felts B, Kelley S, Mahaffy JM, Nulton J, Salamon P, et al. Diversity and population structure of a near-shore marine-sediment viral community. Proc Biol Sci 2004;271(1539):565–74.
25. Lopez-Bueno A, Tamames J, Velazquez D, Moya A, Quesada A, Alcami A. High diversity of the viral community from an Antarctic lake. Science 2009;326(5954):858–61.

26. Beumer A, Robinson JB. A broad-host-range, generalized transducing phage (SN-T) acquires 16S rRNA genes from different genera of bacteria. Appl Environ Microbiol 2005;71(12): 8301–4.

27. Ghosh D, Roy K, Williamson KE, White DC, Wommack KE, Sublette KL, et al. Prevalence of lysogeny among soil bacteria and presence of 16S rRNA and *trzN* genes in viral-community DNA. Appl Environ Microbiol 2008;74(2):495–502.

28. Altschul SF, Gish W, Miller W, Myers EW, Lipman DJ. Basic local alignment search tool. J Mol Biol 1990;215(3):403–10.

29. Kent WJ. BLAT--the BLAST-like alignment tool. Genome Res 2002;12(4):656–64.

30. Lysholm F, Andersson B, Persson B. FAAST: flow-space assisted alignment search tool. BMC Bioinform 2011;12:293.

31. Wommack KE, Bhavsar J, Polson SW, Chen J, Dumas M, Srinivasiah S, et al. VIROME: a standard operating procedure for analysis of viral metagenome sequences. Stand Genomic Sci 2012;6(3):427–39.

32. Kunin V, Copeland A, Lapidus A, Mavromatis K, Hugenholtz P. A bioinformatician's guide to metagenomics. Microbiol Mol Biol Rev 2008;72(4):557–78.

33. Breitbart M, Haynes M, Kelley S, Angly F, Edwards RA, Felts B, et al. Viral diversity and dynamics in an infant gut. Res Microbiol 2008;159(5):367–73.

34. Breitbart M, Hewson I, Felts B, Mahaffy JM, Nulton J, Salamon P, et al. Metagenomic analyses of an uncultured viral community from human feces. J Bacteriol 2003;185(20):6220–3.

35. Kim MS, Park EJ, Roh SW, Bae JW. Diversity and abundance of single-stranded DNA viruses in human feces. Appl Environ Microbiol 2011;77(22):8062–70.

36. Minot S, Grunberg S, Wu GD, Lewis JD, Bushman FD. Hypervariable loci in the human gut virome. Proc Natl Acad Sci USA 2012;109(10):3962–6.

37. Reyes A, Haynes M, Hanson N, Angly FE, Heath AC, Rohwer F, et al. Viruses in the faecal microbiota of monozygotic twins and their mothers. Nature 2010;466(7304):334–8.

38. Zhang T, Breitbart M, Lee WH, Run JQ, Wei CL, Soh SW, et al. RNA viral community in human feces: prevalence of plant pathogenic viruses. PLoS Biol 2006;4(1):e3.

39. Foxman EF, Iwasaki A. Genome-virome interactions: examining the role of common viral infections in complex disease. Nat Rev Microbiol 2011;9(4):254–64.

40. Virgin HW, Wherry EJ, Ahmed R. Redefining chronic viral infection. Cell 2009;138(1):30–50.

41. Patel MR, Emerman M, Malik HS. Paleovirology – ghosts and gifts of viruses past. Curr Opin Virol 2011;1(4):304–9.

42. Feschotte C, Gilbert C. Endogenous viruses: insights into viral evolution and impact on host biology. Nat Rev Genet 2012;13(4):283–96.

43. Lander ES, Linton LM, Birren B, Nusbaum C, Zody MC, Baldwin J, et al. Initial sequencing and analysis of the human genome. Nature 2001;409(6822):860–921.

44. Norman JM, Handley SA, Virgin HW. Kingdom-agnostic metagenomics and the importance of complete characterization of enteric microbial communities. Gastroenterology 2014;146(6):1459–69.

45. Bosch AA, Biesbroek G, Trzcinski K, Sanders EA, Bogaert D. Viral and bacterial interactions in the upper respiratory tract. PLoS Pathogens 2013;9(1):e1003057.

46. Kane M, Case LK, Kopaskie K, Kozlova A, MacDearmid C, Chervonsky AV, et al. Successful transmission of a retrovirus depends on the commensal microbiota. Science 2011;334(6053):245–9.

47. Kuss SK, Best GT, Etheredge CA, Pruijssers AJ, Frierson JM, Hooper LV, et al. Intestinal microbiota promote enteric virus replication and systemic pathogenesis. Science 2011;334(6053):249–52.

48. Robinson CM, Jesudhasan PR, Pfeiffer JK. Bacterial lipopolysaccharide binding enhances virion stability and promotes environmental fitness of an enteric virus. Cell Host Microbe 2014;15(1):36–46.

49. Duerkop BA, Hooper LV. Resident viruses and their interactions with the immune system. Nat Immunol 2013;14(7):654–9.

50. Fichorova RN, Lee Y, Yamamoto HS, Takagi Y, Hayes GR, Goodman RP, et al. Endobiont viruses sensed by the human host – beyond conventional antiparasitic therapy. PloS One 2012;7(11):e48418.

51. Ives A, Ronet C, Prevel F, Ruzzante G, Fuertes-Marraco S, Schutz F, et al. *Leishmania* RNA virus controls the severity of mucocutaneous leishmaniasis. Science 2011;331(6018):775–8.

52. De Vlaminck I, Khush KK, Strehl C, Kohli B, Luikart H, Neff NF, et al. Temporal response of the human virome to immunosuppression and antiviral therapy. Cell 2013;155(5):1178–87.

53. Uhr JW, Dancis J, Franklin EC, Finkelstein MS, Lewis EW. The antibody response to bacteriophage phi-X 174 in newborn premature infants. J Clin Investig 1962;41:1509–13.

54. Eriksson F, Tsagozis P, Lundberg K, Parsa R, Mangsbo SM, Persson MA, et al. Tumor-specific bacteriophages induce tumor destruction through activation of tumor-associated macrophages. J Immunol 2009;182(5):3105–11.

55. Basic M, Keubler LM, Buettner M, Achard M, Breves G, Schroder B, et al. Norovirus triggered microbiota-driven mucosal inflammation in interleukin 10-deficient mice. Inflamm Bowel Dis 2014;20(3):431–43.

56. Cadwell K, Patel KK, Maloney NS, Liu TC, Ng AC, Storer CE, et al. Virus-plus-susceptibility gene interaction determines Crohn's disease gene Atg16L1 phenotypes in intestine. Cell 2010;141(7):1135–45.

57. Young GR, Eksmond U, Salcedo R, Alexopoulou L, Stoye JP, Kassiotis G. Resurrection of endogenous retroviruses in antibody-deficient mice. Nature 2012;491(7426):774–8.

58. Barton ES, White DW, Cathelyn JS, Brett-McClellan KA, Engle M, Diamond MS, et al. Herpesvirus latency confers symbiotic protection from bacterial infection. Nature 2007;447(7142):326–9.

59. White DW, Keppel CR, Schneider SE, Reese TA, Coder J, Payton JE, et al. Latent herpesvirus infection arms NK cells. Blood 2010;115(22):4377–83.

60. Smith KA, Efstathiou S, Cooke A. Murine gammaherpesvirus-68 infection alters self-antigen presentation and type 1 diabetes onset in NOD mice. J Immunol 2007;179(11):7325–33.

61. Nguyen Y, McGuffie BA, Anderson VE, Weinberg JB. Gammaherpesvirus modulation of mouse adenovirus type 1 pathogenesis. Virology 2008;380(2):182–90.

62. Peacock JW, Elsawa SF, Petty CC, Hickey WF, Bost KL. Exacerbation of experimental autoimmune encephalomyelitis in rodents infected with murine gammaherpesvirus-68. Eur J Immunol 2003;33(7):1849–58.

63. Haque A, Rachinel N, Quddus MR, Haque S, Kasper LH, Usherwood E. Co-infection of malaria and gamma-herpesvirus: exacerbated lung inflammation or cross-protection depends on the stage of viral infection. Clinical and experimental immunology 2004;138(3):396–404.

CHAPTER 10

Promises and Prospects of Microbiome Studies

Maria C. Rivera and Jacques Izard

Since Anthony van Leeuwenhoek, first microscopic observations of the unseen microbiota and the more recent realization that little of the microbes in the biosphere are known, humans have developed a deep curiosity to fully understand the inner workings of the microbial realm. Our ability to characterize the complexity of microbial communities in their natural habitats has dramatically improved over the past decade thanks to advances in high-throughput methodologies. By eliminating the need to isolate and culture individual species, metagenomics approaches have removed many of the obstacles that hindered research in the ecology of mixed microbial consortia, providing valuable information about the diversity, composition, function, and metabolic capability of the community.

Microbes are the unseen majority with the capability to colonize every environment, including our bodies. The establishment and composition of a stable human microbiome is determined by the host genetics, immunocompetence, and life-style choices. Our life-style choices determine our exposure to many external and internal environmental factors that permanently or temporarily can influence our microbiome composition. Figure 10.1 illustrates some of the life-style-related factors that might influence the microbiota of the skin, mouth, and gut. It is not limited to what we carry, touch, breath, and eat. Other dispersal vectors include secretion, excretions, aerosols, air flow, animals, moving surfaces, water, beverages, food, contact, wind, tools, toiletry, and others. These influence the microbiome membership, who are present, and they have the ability to participate in the microbiome dynamic within an environment. The establishment of a microbial community is dependent on many environmental factors, including pH, temperature, altitude, weather, soil type, nutrient availability, relative humidity, air quality, pollutants, microbial competitors, and others. In other words, we are superorganisms interconnected with other living forms on this Earth.

Metagenomics for Microbiology. http://dx.doi.org/10.1016/B978-0-12-410472-3.00010-5

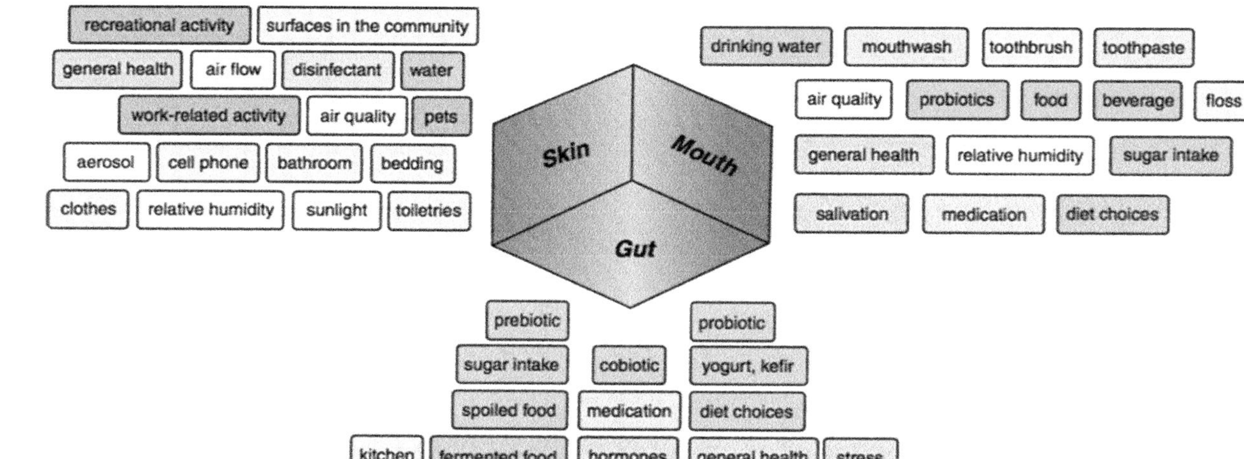

Fig. 10.1. Microbiome of modern humans is influenced by life-style choices. During everyday life, humans are exposed to many external and internal environmental factors that can influence the composition of their microbiome. The human microbiome is here represented by the microbiota of the skin, mouth, and gut, and next to each body site a series of environmental modifiers. Some of these modifiers, such as general health, diet, medications, and stress, can affect the microbiota of more than one body site, but others, such as toothbrush and toothpaste, can directly modify the microbiota of only one body site.

The exploration of the microbial worlds uncovered the extreme microbial diversity throughout the biosphere, from hydrothermal vents on the ocean floor to the intestinal tract of animals. In addition to metagenomics, other technologies including metatranscriptomics, metaproteomics, metabolomics, and metalipidomics can provide better insights of the microbial ecosystems dynamics. Fundamental gaps in knowledge still exist. We will take few examples to discuss the promises of the upcoming discoveries.

MICROBIOME DURING DEVELOPMENT AND DISEASE

Since the 1980s, advances in sequencing technologies have uncovered the immense diversity and functional capabilities of the microbial world. More recently, the introduction of next-generation sequencing (NGS) combined with metagenomics approaches has allowed a better understanding of the complexity of the interactions between animals and their associated microbiota, in particular, the gut microbiome.

The digestive tract of animals has coevolved with a diverse and complex microbial community that responds to host diet and provides metabolic signals to the host during its developmental stages among many other functions.[1] The permeability of the gut facilitates the transport of metabolites produced by the microbiota, which allows the signaling and interactions between the gut microbiota and the host organs and tissues.[2] Host immune and nervous system as well as behavior including mating can be influenced.[1,3,4]

The use of metagenomics approaches to understand the complex metabolic interactions between the human host and its microbiome has revealed extensive variability in the diversity and composition of the microbiome between individuals and throughout the life-span of the host. A representation of this knowledge is summarized in Figure 10.2. Our growing knowledge of the magnitude and complexity of the interactions between the host and the gut microbiome is drastically changing our views of human health, disease, and aging.[2] The lifelong changes in the complexity of the host–microbiome metabolic interactions offer the opportunity for the development of specific therapeutic interventions targeting the gut microbiome throughout the life-span of the host.

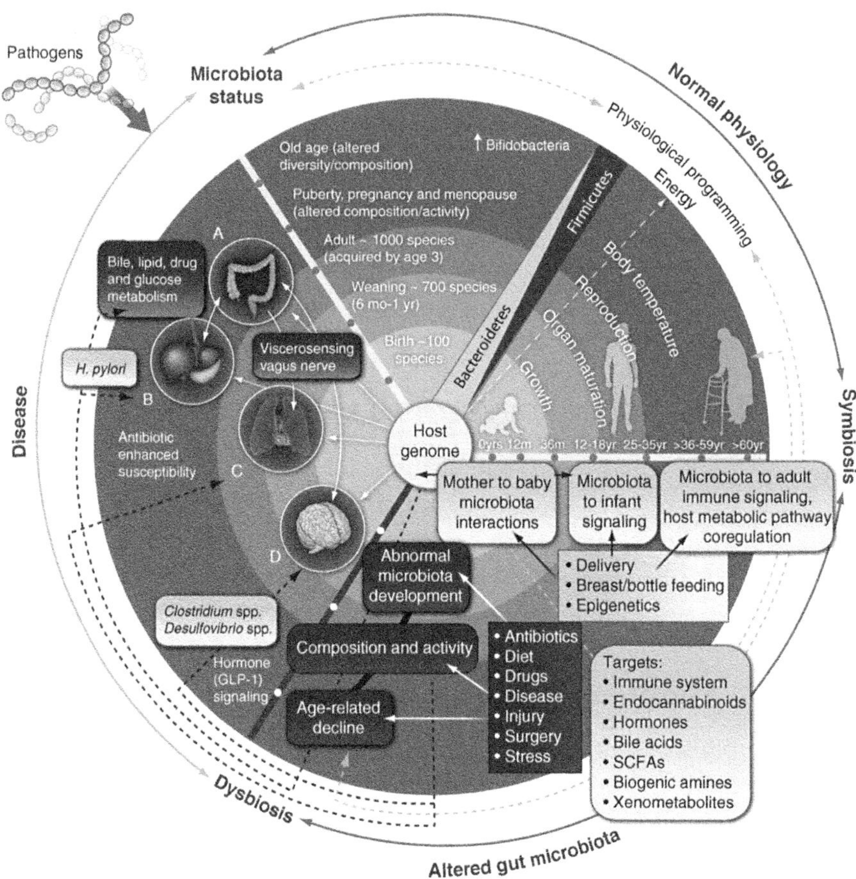

Fig. 10.2. Gut microbiota in development and disease. The influence of the gut microbiota on human health is continuous from birth to old age. The maternal microbiota may influence both the intrauterine environment and the postnatal health of the fetus. At birth, about 100 microbial species populate the colon. Early environmental factors (e.g., method of delivery), nutritional factors (e.g., breastfeeding or bottle feeding), and epigenetic factors have been implicated in the development of a healthy gut and its microbial symbionts. Changes in gut microbial composition in early life can influence risk for developing disease later in life. During suckling, the microbial community develops rapidly; shifts in microbial diversity occur throughout childhood and adult life, and in old age, there is a decrease in the Bacteroidetes *and an increase in the* Firmicutes *species. The gut microbiota is important for maintaining normal physiology and energy production throughout life. Body temperature regulation, reproduction, and tissue growth are energy-dependent processes that may rely in part on gut microbial energy production. Extrinsic environmental factors (such as antibiotic use, diet, stress, disease, and injury) and the mammalian host genome continually influence the diversity and function of the gut microbiota with implications for human health. Disruption of the gut microbiota (dysbiosis) can lead to a variety of different diseases, including (A) inflammatory bowel disease, colon cancer, and irritable bowel syndrome; (B) gastric ulcers, nonalcoholic fatty liver disease, and obesity and metabolic syndromes; (C) asthma, atopy, and hypertension; and (D) mood and behavior through hormone signaling. The gut microbiota is also important for drug metabolism and preventing the establishment of pathogenic microbes.* (Reprinted with permission from the American Association for the Advancement of Science.[2])

THERAPEUTIC POTENTIAL OF MICROBIOTA MODIFICATIONS

Metagenomic studies have uncovered the connections between the disease and the changes in the homeostasis of host–microbiota interactions.[2] Such findings have renewed the interest of the medical community toward therapeutic interventions targeting the gut microbiota. Although unknowingly, humans have performed microbial modification for hundreds of years, the first deliberate use of microbial modification for human health date from the beginning of the last century.[5,6] In 1907, Ellie Metchnioff suggested that the consumption of lactic acid bacteria could improve health and longevity, and initiated the modern probiotics movement. Probiotics are "live microorganisms that, when administered in adequate amounts, confer a health benefit on the host."[5,6] Although the application of comparative genomics approaches have contributed to elucidate the genetic components that confer probiotic properties to certain taxa, metagenomics methodology allows more comprehensive studies of the effect of probiotics.[5,7] Metagenomics together with other omics approaches can potentially identify the metabolites and metabolic pathways inducing the host–microbiota feedback mechanism by which probiotics can modulate health.

Extensive multi-omics studies are needed to fully understand how changes in the ecology of the gut microbiota relate to disease and how to better design and utilize microbial modification therapies.[8] Which are the microbiome structural and homeostatic changes associated with the development of secondary infections following the administration of broad-spectrum antimicrobials?[9] How are the homeostasis and ecological dynamics of the host-microbiome interactions affected by the elimination of specific taxa after immunization? Microbiota modification will soon fall into the realm of personalized and preventive medicine.[10–13]

So far, the best evidence of the therapeutic potential of microbiota modifications is fecal microbiota transplantation (FMT). The replacement of the patient's gut microbiome with the microbiome from a healthy donor has been very successful in the treatment of antibiotic-resistant *Clostridium difficile* infections. More in-depth metagenomics studies from patients with *C. difficile* infections before and after fecal transplants are needed to determine if it is possible to obtain the same therapeutic benefits by introducing selected members of the microbial community or if indeed the full replacement of the gut microbiome is needed.[14,15] The efficacy of therapeutic interventions based on microbiota modulation will have to be demonstrated by their ability to consistently restore a healthy steady state. The success of the first randomized clinical trial demonstrating the effectiveness of the FMT for the treatment of recurrent *C. difficile* infections has catalyzed the formation of FMT-patient advocacy groups, and the establishment of stools banks to ultimately encourage the development of targeted therapies.[14,16,17] The success of FMT suggests that it might be appropriate to establish strict regulations for the screening of infectious disease in microbiome transplants, similar to the ones in place for organ transplants.[14,16,17]

CHALLENGES AHEAD: HUMAN MICROBIOME INFORMATION IN THE HEALTH CARE SYSTEM

In November 2013, the US Food and Drug Administration (FDA) approved the marketing of the NGS systems as diagnostic devices for human genome sequencing.[18,19] The authorization by the FDA of the Illumina sequencing platform for whole human genome surveys in the clinical setting paves the way for the development of new sequencing-based clinical tests, potentially including microbiome profiling. It is predicted that in the future, the patient genome and microbiome data can be integrated allowing the identification of medically relevant variants that might transform clinical research and patient care. Potentially, these combined data sets can help inform about disease predisposition or responses to drugs allowing the design of personalized care and/or early therapeutic interventions. Many challenges need to be conquered before microbiome information is routinely incorporated in health care.

Extensive studies of microbiome profile variations among healthy individuals are needed before microbiome data can be used for predicting disease predisposition, onset, and progression, and drug–response modulation. Similar to the human genome sequencing, the full integration of microbiome data in the clinical setting requires major research efforts in the collection of rigorous evidence supporting the role of the microbiome in health and disease, the development of appropriate regulatory and validation policies, the implementation of policies addressing patients' rights, and the training of the physician and health care professionals in microbiome data interpretation.[20] A lot of research and validation will be required before the approval by the FDA of microbiome profiling in the clinical setting. Accuracy, precision, analytical sensitivity and specificity, reference range, and reportable range will be scrutinized. In the United States, the Clinical Laboratory Improvement Amendments (CLIA), which perform testing of clinical laboratories to ensure their accuracy in testing human samples, will have to develop new standards that can capture the microbiome complexity. As many other technological advances and discoveries, there is likely to be a lapse of time before the actual adoption of the new technology into the clinical setting.

ETHICAL CONSIDERATIONS OF MICROBIOME RESEARCH

In the 21st century, the Internet and other digital technologies have facilitated the access to personal information raising concerns related to privacy and data rights issues. The ease of the data accessibility has serious implications for the regulation of research in human subjects. In microbiome research, those regulatory issues are related to the selection and recruitment of human subjects, the possibility of individual or group stigma associated with the research findings, privacy and confidentiality, and informational risks associated with disclosure of some of the findings.[21–25] Additionally, microbiome research subjects may be identified by the disclosure of information collected from the behaviors survey and/or from the microbiota samples. In addition to the issues related to research-generated data, unique to the human microbiome are issues related to informed consent for future use of stored microbiota samples.[26–28] "Who owns your poop?" is the humorous and thought-provoking question posed by Alice Hawkins and Kieran O'Doherty, while discussing the impact on microbiome research of issues related to privacy, consent, ownership, return of results, governance, and benefit sharing.[28] Although some ethical issues are unique to microbiome research, similar ethical concerns were extensively debated during the Human Genome Project and, more recently, genome-wide association studies.[24,29]

To better address the ethical issues facing contemporary research such as microbiome research, in 2011, the US Department of Health and Human Services issued an Advance Notice of Proposed Rulemaking (ANPRM), entitled "Human Subjects Research Protections: Enhancing Protections for Research Subjects and Reducing Burden, Delay, and Ambiguity for Investigators," proposing changes to the "Common Rule," as the current federal policy for the Protection of Human Subjects is known.[43] Some of the ANPRM proposed changes include: (1) adjusting Institutional Review Board (IRB) review to contemporary research, (2) establishing a single IRB review for multi-institution research, (3) specific written consent for the use of biospecimens, (4) standards for data security and protection of identifiable or potentially identifiable data, (5) implementing a systematic approach to the collection and analysis of data on unanticipated problems and adverse events, and (6) extending the federal rules to apply to all research, regardless of funding source, that is conducted at US institutions that receive some funding from a Common Rule agency for research using human subjects. The proposed revisions to the Common Rule present many challenges to the regulation of research using human subjects deserving immediate attention by the research community.[30]

The ethical issues facing microbiome research require the implementation of novel guidelines on the proper and ethical collection and use of the data generated by this technology. At the same time, to stimulate and facilitate the process of discovery and to keep pace with the latest technological advances, it is imperative for the regulatory agencies to implement a more agile and adaptable evaluation system.

The artist, like the scientist, can expose and highlight the ethical and privacy issues associated with new technologies. To raise awareness of our increasing access to biotechnologies, Gabriel Barcia-Colombo created the art installation 'DNA vending machine'.[31] The installation shows visuals of individuals whom donated mouthwash from which DNA was extracted, as well as a packaged vial of that DNA. The installation is both troubling and natural. Both oral microbiome and human DNA should be present in the beautifully condensed genetic information presented by the artist. One can just wonder how such material can be used, and if we should expect prepackaged microbiomes to be on the supermarket shelves.

Another side of this ethics debate is whether the microbiota has rights. Under the assumption "for the benefit of all," we have the technological capacity to permanently eradiate members of the microbiota or, through synthetic biology, "create new microbiota."[32–34]

CITIZEN SCIENTIST, CROWDSOURCED RESEARCH, AND THE MICROBIOME

The increasing evidence suggesting the important role played by the microbiota in development, aging, and many human diseases has spur great interest from the general public and has introduced the microbiome field to the citizen science movement.[35–37] Using the citizen science principles of crowdsourcing and crowdfunding, companies and open-sourced projects have been established to collect microbiome samples from donors and provide them with a snapshot of their own microbiome profile in exchange for a monetary contribution.[33] Some large projects include the American Gut in association with the Human Food Project and several universities, the Home Microbiome study with the Alfred P. Sloan Foundation and Argonne National Laboratories, and the Project MERCCURI to analyze the space station microbiome. Online resources, such as SciStarter and microBEnet, allow the identification of active studies where individuals can be involved. The popularity and success of these types of projects is attributed to the need of the general public to access scientific information that can potentially impact their lives and health. This need of information is strongly felt in the case of the gut microbiome, as changes in the gut microbiota have been associated with several debilitating human conditions.

The citizen science movement can empower the individuals but at the same time raises several ethical issues and imposes additional responsibilities and concerns to the researchers involved in such projects.[35,38–41] The research community needs to be aware of the implicit expectations of the citizen scientists and making sure that data collection is not the only goal of the project. While moderating the high expectations of the citizen scientist, the researchers need to ensure the proper use of the collected data to help improve the health and knowledge of, for example, the human subjects.[42] Properly used and integrated with other health status data, the generated microbiome profiles can potentially provide information very useful to both the human subjects and the research community. If rigorous studies corroborate the observed associations between disease state and changes in microbiome structure, the information could improve preventive health care and lead to earlier medical interventions.

Although of immense benefit to the research community and to the participants, the modality of data collection by citizen scientist presents serious concerns with privacy and security of the data. Microbiome samples contain not only the individuals' microbiome but also the genetic information that could potentially identify the sample donor. A successful microbiome citizen science project requires the implementation of the best practices for collection, management, security, analysis, and communication of all the data collected.[21–24] It is important to point out the well-known fact that research conducted using data collected from self-selected participants have methodological limitations. Because of selection bias, information bias, and confounding effects, the findings of research using self-selected participants requires cautious interpretation.[40] It requires the implementation of the proper analytical methodology to identify and possibly compensate for those biases and confounding effects.[40] This type of collection efforts require the full awareness of the participants that the observed correlation or associations might lack accuracy or generality because of those methodological limitations.

The use of self-selected participants can potentially restrict the health benefits of microbiome profiling to a small sector of the population, because the data collected is skewed based on socioeconomics, ethnicity, and/or disease state of the donor. Given the current funding situation, it can also limit the testing of alternate hypotheses.[35–41] These data-gathering efforts will facilitate the collection of large number of samples and help the democratization of science by empowering the citizen scientist.

METAGENOMICS, AGRICULTURE, AND FOOD MICROBIOLOGY

For living entities, proper nutrition is a key factor for survival. Agriculture and food microbiology can benefit from metagenomics advances by improving food safety and security, improving the detection of threats to food production and supply, and increasing the productivity of domestic animals and plants. In meat production, the ongoing use of direct-fed microbial could also be optimized in function of the food intake and the animal of choice.

The World Health Organization estimates that worldwide, approximately 2.2 million annual deaths are associated with foodborne and waterborne pathogens, including bacteria, viruses, and parasites.[43] Food and water safety is a worldwide necessity. The depth and high coverage of NGS makes metagenomics a powerful tool in the detection and surveillance of foodborne pathogens, the detection of outbreaks and transmission routes of foodborne diseases, the testing of foods and food-associated environments, and the identification of microbiota that may protect against foodborne illnesses.

Although the use of metagenomics approaches to test foods for the presence of specific foodborne pathogens can be problematic, its application successfully identified the presence of serovars of *Salmonella* in samples that tested negative by bacteriological analytical manual methods and real-time polymerase chain reaction.[43,44] In order to facilitate the use of metagenomics by governmental agencies and the food industry, it is crucial to develop bioinformatics tools tailored to the needs of the food microbiology laboratory and to put in place the appropriate legal and ethical framework for the collection and use of the data generated. The use of metagenomics approaches in the food industry, to better understand, the food-associated microbial communities, can lead to improvements in productivity, quality, and safety of food.[43,45]

Eliminating hunger worldwide, providing desirable food to a larger population, and producing the needed food in a sustainable way are major challenges facing future food security.[46,47] A possible strategy for meeting those challenges in a sustainable way is to increase agricultural yields and production limits by manipulating above–belowground plant interactions with the goal of reducing pests and increasing crop growth. The association between plants and belowground soilborne microbiota increases plant fitness. Plant growth promoting rhizobacteria (PGPR) are able to influence plant growth and increase plant resistance to herbivores and pathogens.[47–49] Metagenomics approaches to uncover the interactions between plants and belowground microbiota in agricultural systems are essential to tailor and target these interactions for maximum benefit to the crops. More in-depth research in the composition and function of the rhizosphere microbiome is needed in order to better understand and identify the activity of PGPR bacteria in different crops. The knowledge obtained by the analyses of rhizosphere microbiome can be used to design PGPR-based interventions that promote plant growth, nutrition, and defense against pests in agricultural systems. The inoculation of soils with beneficial bacteria such as PGPR could be a sustainable approach to increase productions of crops without the input of chemical fertilizers. Also, soil microbiota manipulation can be potentially used to increase nitrogen fixation and reduce the use of fertilizers and subsequent nitrogenization resulting in economical and ecological benefits.[50,51]

SUMMARY

In this overview, we have highlighted some of the promises and prospects of the utilization of omics approaches to the study of microbiomes. Clearly, there is still much work ahead before achieving a comprehensive understanding of the complex dynamics and interactions within the human microbiome and other microbial ecosystems. At this time, microbiome research is moving beyond the identification of genes and/or taxa and toward an emphasis on the application of multi-meta-omics technologies with the goal of sorting out the functions and pathways responsible for the multidirectional interactions between the microbiota, the host, and the environment. A comprehensive understanding of these interactions will contribute to the design of more effective, preventive, and therapeutic interventions. It creates opportunities for a new vision of health and disease treatment that was not imagined in the 20th century.

Many of the resources and technologies required to fulfil the promise of the microbiome are already available. The development of novel bioinformatics frameworks and analytical techniques are essential in order to more efficiently mine and synthesize the information of the so-called "Big Data" embedded in the present research or soon to be generated.

It is interesting to note the concept of scale involved. From the interaction at the atomic scale of the enzyme and its substrate, driving, in part, the functional assembly of a microbiome, to the dynamic interactions of biomes, we have moved from the nanometer to the meter range and beyond. Let's be ready to stretch our minds a little further and create the tools to better handle the rising concepts.

This methodological and conceptual revolution has provided the magnification lens needed to better understand the unseen world in front of our eyes, the microbial communities inhabiting this planet for over 3.5 billion years. It is a scientific revolution that will soon reach every citizen.

ACKNOWLEDGMENTS

We would like to thank Rebecca S. Misra for drawing Figure 10.1. We apologize for limiting the number of references on a vast topic. The work was supported by the National Institutes of Health, under grant CA166150 (Jacques Izard), and an award from the Thomas F. and Kate Miller Jeffress Memorial Trust (Maria C. Rivera).

REFERENCES

1. Kohl KD, Cary TL, Karasov WH, Dearing MD. Restructuring of the amphibian gut microbiota through metamorphosis. Environ Microbiol Rep 2013;5(6):899–903.
2. Nicholson JK, Holmes E, Kinross J, Burcelin R, Gibson G, Jia W, et al. Host-gut microbiota metabolic interactions. Science 2012;336(6086):1262–7.
3. Sharon G, Segal D, Ringo JM, Hefetz A, Zilber-Rosenberg I, Rosenberg E. Commensal bacteria play a role in mating preference of Drosophila melanogaster. Proc Natl Acad Sci USA 2010;107:2005.
4. Jost T, Lacroix C, Braegger CP, Chassard C. New insights in gut microbiota establishment in healthy breast fed neonates. PLoS One 2012;7(8):e44595.
5. Tissier H. Traitement des infections intestinales par la méthode de la flore bactérienne de l'intestin. Crit Rev Soc Biol 1906;(60):359–61.
6. Anukam KC, Reid G. Probiotics: 100 years (1907-2007) after Elie Metchnikoff's Observation. In: Méndez-Vilas A editor. Microbiology Book Series: Communicating Current Research and Educational Topics and Trends in Applied Microbiology. Spain: FORMATEX; 2007. p. 466–74.
7. Veiga P, Gallini CA, Beal C, Michaud M, Delaney ML, DuBois A, et al. *Bifidobacterium animalis* subsp. lactis fermented milk product reduces inflammation by altering a niche for colitogenic microbes. Proc Natl Acad Sci USA 2010;107(42):18132–7.
8. Sokol H, Pigneur B, Watterlot L, Lakhdari O, Bermudez-Humaran LG, Gratadoux JJ, et al. Faecalibacterium prausnitzii is an anti-inflammatory commensal bacterium identified by gut microbiota analysis of Crohn disease patients. Proc Natl Acad Sci USA 2008;105(43):16731–6.
9. Antonopoulos DA, Huse SM, Morrison HG, Schmidt TM, Sogin ML, Young VB. Reproducible community dynamics of the gastrointestinal microbiota following antibiotic perturbation. Infect Immun 2009;77(6):2367–75.
10. Guyonnet D, Chassany O, Ducrotte P, Picard C, Mouret M, Mercier CH, et al. Effect of a fermented milk containing *Bifidobacterium animalis* DN-173 010 on the health-related quality of life and symptoms in irritable bowel syndrome in adults in primary care: a multicentre, randomized, double-blind, controlled trial. Aliment Pharmacol Ther 2007;26(3):475–86.
11. Kekkonen RA, Lummela N, Karjalainen H, Latvala S, Tynkkynen S, Jarvenpaa S, et al. Probiotic intervention has strain-specific anti-inflammatory effects in healthy adults. World J Gastroenterol 2008;14(13):2029–36.

12. Lemon KP, Armitage GC, Relman DA, Fischbach MA. Microbiota-targeted therapies: an ecological perspective. Sci Transl Med 2012;4(137):137rv5.

13. Licht TR, Hansen M, Bergstrom A, Poulsen M, Krath BN, Markowski J, et al. Effects of apples and specific apple components on the cecal environment of conventional rats: role of apple pectin. BMC Microbiol 2010;10:13.

14. Cammarota G, Ianiro G, Gasbarrini A. Fecal microbiota transplantation for the treatment of *Clostridium difficile* infection: a systematic review. J Clin Gastroenterol 2014;48(8):693–702.

15. Petrof EO, Gloor GB, Vanner SJ, Weese SJ, Carter D, Daigneault MC, et al. Stool substitute transplant therapy for the eradication of *Clostridium difficile* infection: "RePOOPulating" the gut. Microbiome 2013;1(1):3.

16. Owens C, Broussard E, Surawicz C. Fecal microbiota transplantation and donor standardization. Trends Microbiol 2013;21(9):443–5.

17. Smith MB, Kelly C, Alm EJ. Policy: How to regulate faecal transplants. Nature 2014; 506(7488):290–1.

18. FDA Press Release N. FDA allows marketing of four "next generation" gene sequencing devices, 2013.

19. Collins FS, Hamburg MA. First FDA authorization for next-generation sequencer. N Engl J Med 2013;369(25):2369–71.

20. Gargis AS, Kalman L, Berry MW, Bick DP, Dimmock DP, Hambuch T, et al. Assuring the quality of next-generation sequencing in clinical laboratory practice. Nat Biotechnol 2012;30(11):1033–6.

21. Wolf SM, Crock BN, Van Ness B, Lawrenz F, Kahn JP, Beskow LM, et al. Managing incidental findings and research results in genomic research involving biobanks and archived data sets. Genet Med 2012;14(4):361–84.

22. Human Microbiome Project Consortium. A framework for human microbiome research. *Nature* 2012; 486(7402): 215–21.

23. McGuire AL, Colgrove J, Whitney SN, Diaz CM, Bustillos D, Versalovic J. Ethical, legal, and social considerations in conducting the Human Microbiome Project. Genome Res 2008;18(12):1861–4.

24. Hoffmann DE, Fortenberry JD, Ravel J. Are changes to the common rule necessary to address evolving areas of research? A case study focusing on the human microbiome project. J Law Med Ethics 2013;41(2):454–69.

25. Tryka Ka, Hao L, Sturcke A, Jin Y, Wang ZY, Ziyabari L, et al. NCBI's Database of Genotypes and Phenotypes: dbGaP. Nucleic Acids Res 2014;42:D975–9.

26. Slashinski MJ, McCurdy SA, Achenbaum LS, Whitney SN, McGuire AL. "Snake-oil," "quack medicine," and "industrially cultured organisms": biovalue and the commercialization of human microbiome research. BMC Med Ethics 2012;13:28.

27. Master Z, Nelson E, Murdoch B, Caulfield T. Biobanks, consent and claims of consensus. Nat Methods 2012;9(9):885–8.

28. Hawkins AK, O'Doherty KC. "Who owns your poop?": insights regarding the intersection of human microbiome research and the ELSI aspects of biobanking and related studies. BMC Med Genomics 2011;4:72.

29. Lewis C, Hill M, Skirton H, Chitty LS. Non-invasive prenatal diagnosis for fetal sex determination: benefits and disadvantages from the service users' perspective. Eur J Human Genet 2012;20:1127–33.

30. Henry LM. Introduction: revising the common rule: prospects and challenges. J Law, Med Ethics 2013;41:386–9.

31. Barcia-Colombo G. DNA Vending Machine. 2014. (Accessed April 26, 2014) http://www.gabebc.com/DNA-Vending-Machine.

32. Cockell CS. The Microbial Stages of Humanity. Interdiscip Sci Rev 2011;36:301–13.

33. Riis S. The ultimate technology: the end of technology and the task of nature. Artif Life 2011;19:471–85.

34. Smith HO, Hutchison CA 3rd, Pfannkoch C, Venter JC. Generating a synthetic genome by whole genome assembly: phiX174 bacteriophage from synthetic oligonucleotides. Proc Natl Acad Sci USA 2003;100(26):15440–5.

35. Riesch H, Potter C. Citizen science as seen by scientists: Methodological, epistemological and ethical dimensions. Public Underst Sci 2014;23(1):107–20.

36. Crall AW, Jordan R, Holfelder K, Newman GJ, Graham J, Waller DM. The impacts of an invasive species citizen science training program on participant attitudes, behavior, and science literacy. Public Underst Sci 2013;22(6):745–64.

37. Freitag A, Pfeffer MJ. Process, not product: investigating recommendations for improving citizen science "success". PLoS One 2013;8(5):e64079.

38. Cooper CB. Is there a weekend bias in clutch-initiation dates from citizen science? Implications for studies of avian breeding phenology. Int J Biometeorol 2013;58(7):1415–9.

39. Kaye J, Curren L, Anderson N, Edwards K, Fullerton SM, Kanellopoulou N, et al. From patients to partners: participant-centric initiatives in biomedical research. Nat Rev Genet 2012;13(5):371–6.

40. Janssens AC, Kraft P. Research conducted using data obtained through online communities: ethical implications of methodological limitations. PLoS Med 2012;9(10):e1001328.

41. McGuire AL, Achenbaum LS, Whitney SN, Slashinski MJ, Versalovic J, Keitel WA, et al. Perspectives on human microbiome research ethics. J Empir Res Hum Res Ethics 2012;7(3):1–14.

42. Berkman B, Hull S, Eckstein L. The Unintended Implications of Blurring the Line between Research and Clinical Care in a Genomic Age. Personal Med 2014;11(3):285–95.

43. Bergholz TM, Moreno Switt AI, Wiedmann M. Omics approaches in food safety: fulfilling the promise? Trends Microbiol 2014;22(5):275–81.

44. Lukjancenko O, Wassenaar TM, Ussery DW. Comparison of 61 sequenced *Escherichia coli* genomes. Microb Ecol 2010;60(4):708–20.

45. Ercolini D. High-throughput sequencing and metagenomics: moving forward in the culture-independent analysis of food microbial ecology. Appl Environ Microbiol 2013;79:3148–55.

46. Godfray HCJ, Beddington JR, Crute IR, Haddad L, Lawrence D, Muir JF, et al. Food security: the challenge of feeding 9 billion people. Science 2010;327:812–8.

47. Orrell P, Bennett AE. How can we exploit above-belowground interactions to assist in addressing the challenges of food security? Front Plant Sci 2013;4:432.

48. Beneduzi A, Ambrosini A, Passaglia LMP. Plant growth-promoting rhizobacteria (PGPR): Their potential as antagonists and biocontrol agents. Genet Mol Biol 2012;35:1044–51.

49. Pangesti N, Pineda A, Pieterse CM, Dicke M, van Loon JJ. Two-way plant mediated interactions between root-associated microbes and insects: from ecology to mechanisms. Front Plant Sci 2013;4:414.

50. Lugtenberg B, Kamilova F. Plant-growth-promoting rhizobacteria. Annu Rev Microbiol 2009;63:541–56.

51. Mapelli F, Marasco R, Rolli E, Barbato M, Cherif H, Guesmi A, et al. Potential for plant growth promotion of rhizobacteria associated with Salicornia growing in Tunisian hypersaline soils. Biomed Res Int 2013;2013:248078.

Brief Glossary of Terms Used in Metagenomics

These are concise definitions to help you understand some of the concepts at play. Resources are available to deepen your understanding from environmental microbiology, medical microbiology, experimental design, taxonomy, study design, statistics, ecology, bioinformatics, etc.

Alpha diversity Expression of microbial diversity at one site taking into account the number of species and their abundance. A community will have a high alpha diversity when there is a high number of species in similar relative abundance.

Aliasing Phenomenon that renders time-varying signals indistinguishable when they are sampled below the oscillation frequencies of the signals.

Autocorrelation plot Method of analyzing time-varying dependencies, in which the correlation of a time-series with successively lagged versions of itself is plotted.

Automated experimental design Statistical technique in which an algorithm is used to generate a design for future experiments, such as the time-points at which to sample subjects in a longitudinal study, based on certain information and objectives provided by the user.

Autoregressive model Statistical model in which the present value of a variable is expressed as a function of its past values.

Bayesian model Type of statistical model, in which uncertainty in knowledge about a system takes into account both observed data as well as prior information or beliefs.

Beta diversity Expression of the difference between sites. It takes into account the nestedness (maintenance of a subset of the population or the whole), and the turnover (species replacement), while being influenced by species richness.

Binning Process of clustering sequences based on their nucleotide composition. This process can be enhanced using reference sequences.

Biofilm Assemblage of microorganisms associated to themselves and/or to a substrate.

Chimera Artificial DNA sequence created during enzymatic amplification and/or data processing.

Collector's curve A plot of the cumulative number of species recorded as a function of the sampling effort and/or sequencing depth. It can then be extrapolated to estimate the species richness.

Contig Contiguous stretch of a DNA sequence comprised of a set of overlapping DNA segments (reads).

Clustering algorithm Computational method for separating data into cohesive groups according to some implicitly or explicitly specified metric.

Diploidy Existence of two complete genome copies in one cell.

Dirichlet process Type of stochastic process often used as a prior probability distribution in infinite mixture models.

Ecological statistics Set of statistical tools developed and validated for macroecology that can be applied to microbiome ecology.

Gene conversion Genetic phenomenon characterized by the nonreciprocal genetic exchange between homologous sequences.

Hidden Markov model Type of state-space model that assumes a discrete set of states with transitions between states occurring at discrete time-steps; states are assumed to be latent or hidden variables that are inferred from data.

Infinite mixture model Bayesian nonparametric clustering technique, in which observations are modeled as arising from a mixture with an infinite number of clusters or components.

Information theory Mathematical theory quantifying the amount of information in data, often involving the measure of entropy, which quantifies the degree of uncertainty involved in predicting the value of a random variable or outcome of an experiment.

k-mer A k-mer is a DNA (or amino acid) substring of length *k*. The occurrence of all possible substrings can be used as a k-mer frequency analysis to characterize and empirically compare genomic sequences. The optimal length k typically varies between 3 and 100 nucleotides based on the application.

Lowest common ancestor Taxonomic or phylogenetic strategy that identifies the root of the smallest subtree of the taxonomy or phylogeny that includes all the genomes of interest.

Machine learning Branch of artificial intelligence aimed at prediction based on patterns learned from a given input dataset, also called training dataset.

Mate pairs Read sequences known to be in 3′ and 5′ ends of a contig. The reads may or may not overlap.

Merooligoploidy Unequal genome dosage during cell division. When the cell division time becomes shorter than the time to replicate and segregate the chromosome, bacteria start a new round of DNA replication before the previous round has been terminated; thus, the gene dosage of regions near the replication origin becomes higher than of regions near the terminus.

Metadata Collected information regarding the experimental parameters, as well as additional information enhancing the study at hand.

Metabolomics Study of small molecules from microbial community, tissue, or complex sample (metabolome).

Metagenomics Study of the genetic information coded by the DNA from microbial community, tissue, or complex sample (metagenome).

Metaproteomics Study of the proteins of a microbial community, tissue, or complex sample **(proteome)**.

Metatranscriptomics Study of the genetic information coded by the RNA from microbial community, tissue, or complex sample (transcriptome).

Microbiome The community composed of microorganisms in a defined habitat. Those organisms encompass bacteria, archaea, lower eukaryotes, phages, and viruses.

Monoploidy The bacteria have a single genome copy per cell, located on a single or multiple genetic elements.

Nonparametric model Type of statistical model in which the structure of the model is not fixed, and that typically grows in size to match the complexity of data.

Oligoploidy The bacteria have between 2 and 10 genome copies in one cell, located on a single or multiple genetic elements.

OTU Operational taxonomic unit is usually defined by 16S rDNA similarities. The similarity percent chosen is often 97% for short reads and 98.5% for full-length reads. It is used in place of taxonomic organization such as species, genus, and others. This information can later be overlaid.

Paired ends Sequence at both ends (5′ and 3′) of a segment of DNA during a sequencing process that reads from both extremities of the same DNA fragment.

Phylotype Observed similarity that classifies a group of organism together. It is rank neutral, thus it can be used for a family, genus, species, or strain based on similarities.

Polyploidy The bacteria have more than 10 genome copies per cell, located on a single or multiple genetic elements.

Q20 or Q30 See quality score.

Quality score, Q score, or Phred quality score It is a parameter used to inform about the base calling accuracy. It is a property that is logarithmically related to the base calling error probability. For example, a Q score of 20 (Q20) to a base is equivalent to the probability of an incorrect base call of 1 in 100 times (or 99% base accuracy). The higher the score is, the greater the accuracy.

Rarefaction curve It allows the estimation of the number of species in case of a reduced sampling effort, to facilitate the comparison between communities with unequal sampling efforts.

Regression model Statistical model in which the value of a dependent variable is expressed as a function of a set of independent variables (covariates).

Scaffold When paired reads are present in two different contigs, the two contigs can be linked to form a scaffold, which is a larger noncontiguous

DNA sequence. The gap size between the two contigs is estimated based on the insert size of the read pairs.

Sequence assembly Process of putting the DNA fragments (sequence reads) back together based on sequence similarity between reads to reconstruct the original sequence.

Sequencing coverage Mean number of times a nucleotide is sequenced in a genome (sequencing depth). Greater coverage will minimize errors due to sequencing or assembly.

Spline model Statistical model used for regression analyses, which is specified piecewise as a series of polynomial functions that are constrained to join together smoothly.

State-space model Statistical model that assumes the outputs or measurements of a system depend on its state, which can change over time.

Taxonomic profiling Identification of the organisms present in a microbial community and their relative abundances.

Whole-metagenome shotgun sequencing In the same way as whole-genome shotgun sequencing refers to short-read sequencing of the complete genome of a single organism, whole-metagenome shotgun sequencing refers to short-read sequencing of the entire mixture of all organisms in a microbiome sample (whole-community sequencing).

Xenobiotic Foreign chemical substance encountered by an organism. This can include pharmaceutical drugs, dietary compounds, or pollutants such as dioxins.

Now it's time to have your feet dangling at the edge of leading science!

INDEX

I

K

L

M

Printed and bound by CPI Group (UK) Ltd, Croydon, CR0 4YY
03/10/2024
01040421-0016